Macroscopic Superconducting Phenomena

An interactive guide

Macroscopic Superconducting Phenomena

An interactive guide

Antonio Badía-Majós

Department of Condensed Matter Physics, University of Zaragoza, Zaragoza, Spain
and
Aragón Nanoscience and Materials Institute (INMA), CSIC-University of Zaragoza,
Zaragoza, Spain

IOP Publishing, Bristol, UK

ISBN   978-0-7503-2711-4 (ebook)
ISBN   978-0-7503-2709-1 (print)
ISBN   978-0-7503-2712-1 (myPrint)
ISBN   978-0-7503-2710-7 (mobi)

DOI   10.1088/978-0-7503-2711-4

Version: 20211101

IOP ebooks

British Library Cataloguing-in-Publication Data: A catalogue record for this book is available from the British Library.

Published by IOP Publishing, wholly owned by The Institute of Physics, London

IOP Publishing, Temple Circus, Temple Way, Bristol, BS1 6HG, UK

US Office: IOP Publishing, Inc., 190 North Independence Mall West, Suite 601, Philadelphia, PA 19106, USA

MATLAB is a registered trademarks of The MathWorks, Inc. See mathworks.com/trademarks for a list of additional trademarks.

Supplementary material is available for this book from http://iopscience.iop.org/book/978-0-7503-2711-4.

To my dearest family, for being so patient, supportive and encouraging throughout the writing of this book

Contents

Appendices

Preface

Superconductivity is one of the most fascinating topics in condensed matter physics, and probably in physics. More than one century after its discovery it still offers a challenging battlefield for condensed matter physicists, who are struggling to find the rationale for a yet growing list of novel experimental facts. A number of spectacular manifestations of the phenomenon (such as zero electrical resistance, the expulsion of magnetic fields or the so-called Josephson effect) have also stimulated the imagination of engineers, thus allowing to conceive and develop a number of applications already present in our society. Just to mention a few, recall that *superconducting coils* are worldwide used in magnetic resonance instruments, magnetic levitation trains, particle accelerators, etc and that ultrasensitive magnetic flux detection based on the so-called *SQUID magnetometers* is already a standard technique in research laboratories.

Flashing back to the pathbreaking statement by Fritz London[1] that 'Superconductivity is a quantum mechanism of macroscopic scale' it is apparent that very different degrees of approximation may be adopted in the study of superconducting phenomena. Thus, the point of view could range from the investigation of coupling mechanisms between electrons at the atomic level to the engineering models used for the design of superconducting MRI coils. Quite closer to the latter, this book puts focus on the macroscopic side. More technically, we adopt the thermodynamic point of view that the properties of interest rely on statistical averages that smear out the microscopic fluctuations. Predominantly, the physical quantities of interest will be the electromagnetic fields averaged over distances much above the fundamental characteristic lengths of the material (typically at the nanometre scale). Nevertheless, in order to ensure a solid basis for the phenomenological macroscopic material laws that will supplement the Maxwell equations, some elementary notions of the microscopic facts will be issued.

Undeniably, a good number of excellent textbooks, already available for decades, are close at hand and cover most of the fundamental aspects of the topic at several degrees of approximation. In this sense, the very purpose of our exposition of the 'theoretical background' is just to offer a somehow personal and dedicated point of view aiming at the application of the basic concepts through a number of selected problems.

Some fundamental topics are not even touched because they are not of (from my view) direct application in the practical section of the book (part III), but of clear importance in the development of superconductivity: intermediate state, thermodynamic properties, the behaviour at high frequencies, the microscopic theory, and so on.

Obviously, in not so few occasions, and because superconductivity is a rather active area, one is led to explore a vast collection of research papers and go through some specific publication. In this regard, I want to stress that it was not my focus to

---

[1] London F 1950 *Superfluids* (New York: Wiley)

compile an exhaustive list of articles related to each topic. The literature provided is just a selection of suitable works that are accessible and may be useful for the potential reader.

In this book, our primary goal is to provide a practical guide for the newcomer in the field of superconductivity, and on occasion specific utilities for the specialist, with emphasis on *computer based* interactive material. The reader will be led through basic physical properties by a number of 'interactives'. Essentially, this means that, following a brief exposition of selected physical phenomena, the reader is provided with a set of ready to use tools that give hands-on experience on such topics as vanishing resistivity, magnetic flux expulsion, critical currents, flux quantisation and others. Interactive software should contribute to the self-learning process and the reader will likely gain the ability to analyse situations beyond the scope of classical analytical methods. Taking further advantage of this, we will introduce a number of resources related to superconducting applications such as levitation devices and magnetic invisibility. Emphasis will be made on the interpretation of widespread experimental studies aimed at the characterisation of superconducting parameters such as the London penetration depth or the critical current density.

The theory will be exposed at a level that targets the minimum complication, but without loss of utility. This means that physical models will be built with the least and simplest mathematical elements that may provide a useful representation of the real facts under study. Keeping in mind the ultimate goal of producing a useful description of the physical system, our emphasis will be to provide minimal tools that give a consistent description of the facts. Along this line, we will insist that a number of experimental issues may be reasonably described by one-dimensional (1D) or quasi-1D models, generally solvable by means of analytic expressions and a simple calculator. Nevertheless, the reader will be warned that many realistic situations demand more elaborate models (i.e. at least 2D) if one wants to go quantitative. Already at this level numerical evaluations may not be skipped. Still, we will show that problem solving may be accomplished by the non-specialist in numerical calculus. At the time being, the widespread availability of computers or equivalent platforms opens a wide range of possibilities much further than conventional analytical tools. Thus, taking benefit of some popular high level software facilities, one has the opportunity of obtaining rather accurate quantitative predictions, with a minimum knowledge of numerical techniques. To be specific, the book puts forward a compilation of computer programs that simulate well-known experimental situations aided by a few lines of code. The programs are built in such a way that straightforward hands-on operation is allowed. Being supplied with an open source, the reader will also be able to self-upgrade the programs with ease so as to investigate new situations and fulfil a particular curiosity or specific needs. For example, it is well known that the levitation forces between a magnet and a superconductor are highly dependent on their geometry and also on the manipulation process (cooling path, etc). The reader will be able to straightforwardly use (and probably optimise) the software so as to check these properties. Indeed, having the focus on the pedagogical value of the sources, here and there we have not implemented the most advantageous solutions from the computational point of view.

Guided by my personal experience in the field, and inspired by unvaluable comments of many colleagues, I have selected a number of situations that frequently require a non-trivial analysis if one wants to connect the 'theoretical' textbook concepts and the real observations in experiments. Specifically, the book focuses on: (i) the influence of demagnetising effects in magnetic measurements, (ii) the analysis of resistive transitions, (iii) the interpretation of magneto-optical imaging of superconducting films, (iv) the study of force microscopies and (v) magnetic levitation systems. Also, a set of elementary tools designed for getting familiar with the behaviour of Josephson junctions are provided.

In principle, the theoretical background needed for a comprehensive utilisation of the book is contained in the syllabus of graduate-level students of physics and engineering. Nevertheless, so as to unify style and notation, I introduce some reference chapters on electromagnetic theory and mathematical issues. Furthermore, in order to provide a self-contained material, some notes are included on the software. For each application, the reader is provided with resources in the form of MATLAB (and OCTAVE compatible) codes. Though not claiming that this is the only or best choice, it is my experience that it is a very good one. In general, with scarce technical difficulties one may perform powerful numerics in a modest personal computer[2].

September 2021, Zaragoza, Spain

---

[2] I have to express my gratitude to the late Prof. E H Brandt, our reference in so many topics in superconductivity. In particular, he prompted us with a repeated question: 'why don't you use matlab?'.

How to use this book

- The book is divided in three parts. Part I is devoted to introducing the fundamental theoretical concepts, as well as to review a number of celebrated experiments along the course of superconducting history. Part II is put forward as a basic background in mathematical and computational concepts that furnish the reader with tools for practice and application. Finally, part III presents a collection of 'classical' problems of macroscopic superconductivity. Each problem (one per chapter) has been analysed with the help of numerical resources that are supplied. All the programs are written from scratch in high level language, and thus may be manipulated (and probably improved) by the reader.

- Interactivity has been extended to the majority of the contents. Nearly every figure and related equations may be worked out with the help of the accompanying codes. For the readers' sake they have been documented internally with references to the main text. Availability of codes has been flagged with the $\mathscr{O}$ symbol aside the figures, that is a link to the related software.

- The actual position of part II is somehow flexible. It is the reader themself who could decide either to proceed with my suggestion, change order or even consider it as an appendix, depending on their background or particular interest. For the real newcomer to superconductivity who also wants an update in basic calculations, I would recommend following the proposed order: start with part I, continue with part II, and eventually jump to part III, perhaps with some revision of part I. This means, first get motivation on the kind of physical properties that are targeted, then learn the tools for modelling and eventually go to specific calculations.

- Owing to its nature, this book is designed to be read by electronic means. Internal and external navigation are available and concurrent practice with the software (either with resident software or in the cloud) is strongly recommended.

Acknowledgements

This book gathers my insights into some problems in the field of superconductivity. They are greatly due to the very inspiring feedback of many colleagues and friends whose collaboration is to be acknowledged.

I am especially indebted to Prof. R Navarro and Prof. C Rillo who introduced me in the field at the very dawn of high temperature superconductivity. Also, I must express my gratitude to Dr C López (Carlos) and Dr J L Giordano (Luis) my colleagues and friends, whose enthusiastic discussions have illuminated the path over the years.

It is a pleasure to recall a number of visits that have allowed me to dive into a collection of thrilling topics. As a visitor of the Low Temperature group at Centro Atómico de Bariloche I had the chance to learn about the physics of resistive transitions. Many thanks are due to the dedication and hospitality of Dr G Nieva, Dr J Guimpel and Dr F Luzuriaga. During my visit to the Ames Lab, it was my privilege to be hosted and instructed by Prof. J R Clem about the theory of the critical state. I am also grateful to Prof. L Prigozhin (Ben Gurion University) for clarifying discussions about numerical techniques in applied superconductivity. While visiting the École Polytechnique, I had the chance to learn about magneto-optical imaging from Prof. C J van der Beek and his group (a special thought for Dr M Grisolia). I am also grateful to Prof. P Grutter's group from McGill University with whom I could learn about magnetic force microscopy. In the course of my postdoctoral period in the group of Prof. H C Freyhadt (University of Göttingen) I learned about magnetic levitation with superconductors, a topic that has fascinated me ever since. Currently, a fruitful collaboration with Prof. K Öztürk and his group (Karadeniz Technical University) is allowing me to improve the knowledge on the practical issues of this subject. Finally, I would also like to thank Dr Y Genenko and Dr S Yampolskii for introducing me to the thrilling topic of magnetic invisibility during my visit to Darmstadt University. Many thanks to all.

Very special thanks are due to Prof. L A Angurel. Not only have I to recall so many hours deep in discussions about superconductivity but also acknowledge his support and infinite patience during the writing this book, dispensing me from any other duties.

Zaragoza, Spain
September 2021

Author biography

Antonio Badía-Majós

Antonio Badía-Majós obtained his doctoral degree from the University of Zaragoza (Spain) in 1993. Following a postdoctoral period at the University of Göttingen (Germany), he returned and became Associate Professor for the Condensed Matter Physics Department of the University of Zaragoza, and also a Fellow of the Material Science Institute of Aragón (INMA at present).

Along his career he has been mainly devoted to the Physics education of engineering and mathematics students, authoring several handbooks and publications in taught physics.

Concerning the research activities, they have been focused on several aspects of superconductivity, with contributed papers in topics ranging from experimental work to phenomenological theories, numerical simulations, and microscopic aspects. A predominant interest is the application of mathematical physics methods to the modelling of electromagnetic properties of superconductors.

Detailed information at: http://personal.unizar.es/anabadia/index_en.html

Part I

Physical facts: theory and experiments

IOP Publishing

Macroscopic Superconducting Phenomena
An interactive guide
Antonio Badía-Majós

Chapter 1

Elements of electromagnetic theory

As related to the main focus of this book, in this chapter we will briefly review some key concepts on the *classical electromagnetic theory*. Basically, as well as recalling a minimal conceptual framework, we aim at providing a unifying point of view as concerns the specific applications considered in part III. In particular, this affects the not infrequent confusion related to *convention* related aspects when one compares different bibliographical sources[1]. Also, we will introduce a number of ideas 'less conventional' in standard physics courses. Thus, as related to a topic of high relevance in macroscopic superconductivity, i.e. the appearance of resistive transient states, we will introduce some basic ideas about the coupling between electromagnetic and thermodynamic energy modes in matter. In particular, we will discuss the so-called principle of *minimum entropy production*, which will be the basis for further modelling.

Certainly, a good number of excellent textbooks treating classical electromagnetism are available in most libraries, and we could just address the reader to a number of them. The target of this chapter is not covering the classical topics, but just to offer a selection of dedicated concepts. Concerning some reference for further background, we will just mention two of our favourite books: Jackson's *Classical Electrodynamics* [1] and Griffiths' *Introduction to Electrodynamics* [2]. They are just as valid as many others.

Being focused on the behaviour of superconducting materials in the low frequency regime, we will be concerned with magnetostatics ($\dot{\mathbf{B}} = 0$) or at most with magneto-quasi-statics ($\dot{\mathbf{B}}$ small). Below, we will define such conditions for electromagnetic systems and introduce the quantities of interest.

---

[1] For instance, as concerns the usage of $\mathbf{B}$, $\mathbf{b}$, $\mu_0\mathbf{H}$ or $\mu_0\mathbf{h}$ for the magnetic field or the concept of 'magnetisation' $\mathbf{M}$ for superconductors.

doi:10.1088/978-0-7503-2711-4ch1

1.1 Low frequency electrodynamics: fundamentals

Since the early days of superconductivity[2], it has been accepted that electrostatic fields have a negligible influence on the electromagnetic properties of these materials. Although some recent investigations suggest that this topic should be reconsidered at least in some microscopic phenomena, we will adopt the 'classical' ansatz that, same as in metallic conductors, surface screening gives way to a negligible influence of electrostatic fields.

As concerns the physical modelling of time-dependent phenomena, our focus will be on transitions that occur slowly enough so as to allow current recombination with fleeting (and negligible in observations) presence of charge accumulation. In other words, the relevant mechanism of electromagnetic energy storage will be in the form of electric currents. In practice, this may occur for conducting materials with dimensions in the macroscopic scale. As explained in [4] electric currents are the predominant mechanism for good conductors, as long as it may verified by checking for typical values of σ in the absence of either dielectric or magnetic response:

$$\sigma > \frac{1}{\ell}\sqrt{\frac{\epsilon_0}{\mu_0}} \tag{1.1}$$

with ℓ being the typical smallest dimension of the system. At first glance, super-conductors ($\sigma \to \infty$) safely fulfil this inequality (charges are extremely mobile). However, as the above criterion was deduced for nondispersive ohmic media, one should go a bit deeper into the microscopic theory of superconductivity to be sure that things work so. In fact, they do [4] and we can add that even microwave superconducting devices are designed under this framework.

1.1.1 Maxwell equations

According to the above arguments, our quantities of interest will be the fields **B** (magnetic field), **E** (electric field) and **J** (current density). Below, we will discuss how to incorporate them to the physical laws that will describe the targeted macroscopic superconducting phenomena.

Field equations. To start with, we put forward the Maxwell equations in the 'simplified' form that corresponds to our **quasi-stationary** approach[3]

$$\nabla \times \mathbf{B} = \mu_0 \mathbf{J} \qquad \text{(Ampere's law)}$$

$$\nabla \times \mathbf{E} = -\frac{\partial \mathbf{B}}{\partial t} \qquad \text{(Faraday's law)} \tag{1.2}$$

$$\nabla \cdot \mathbf{B} = 0 \qquad \text{(solenoidality of } \mathbf{B}\text{)}$$

---

[2] The topic of electrostatic fields in superconductors was considered by the London brothers and searched in a dedicated experiment [3]. It was concluded that electrostatic effects are undetectable.

[3] In the literature, one may find alternative names for this situation: (quasi)static, (quasi)steady and (quasi) stationary. The first one focuses on the magnetic field while the two latter refer to its sources, i.e. the moving charges. Thus, a steady (or stationary) current produces a magnetostatic field.

Additionally, the following considerations are due:

- Related to the low frequency condition, consistently with the divergence of Ampère's law, the continuity equation reads $\nabla \cdot \mathbf{J} \approx 0$.
- Concerning the divergence of the electric field $\nabla \cdot \mathbf{E}$, as shown above, we will not use it as a straightforward source equation, i.e. $\nabla \cdot \mathbf{E} = \rho_q/\epsilon_0$. When required, it will be specified subsequent to the material law that properly describes the conductor[4].
- In numerical calculations, one usually implements Biot–Savart's law instead of recalling Ampère (and solenoidality follows):

$$\mathbf{B}(\mathbf{r}) = \frac{\mu_0}{4\pi} \int \frac{\mathbf{J}(\mathbf{r}') \times (\mathbf{r} - \mathbf{r}')}{\|\mathbf{r} - \mathbf{r}'\|^3} \, dV' \qquad (1.3)$$

Thus, in the general statement, we will have three fundamental vector quantities, whose determination requires nine consistent scalar equations. The main issue will be to find the correct law for the superconducting material, as it will be later discussed. By now, let us complete the general framework.

Electromagnetic potentials. As it is customary in many problems in electromagnetics, the introduction of potentials may be very useful from the mathematical point of view so as to prompt the solution of the involved differential equations. No less important is that potentials may be the link connecting theory and experiment. Just recall that what is measured in the laboratory are potential differences, and not electric fields. In macroscopic superconductivity, one frequently works with magnetic potential:

$$\mathbf{B} = \nabla \times \mathbf{A} \qquad (1.4)$$

and occasionally with the relation

$$\mathbf{E} = -\frac{\partial \mathbf{A}}{\partial t} - \nabla\phi \qquad (1.5)$$

$\nabla\phi$ being the electric (scalar) potential.

The reader may wonder about the consistency of equation (1.5) with our previous statement that electrostatic fields may be neglected. Why should one use the electrostatic potential term $\nabla\phi$ instead of plainly writing the following?

$$\mathbf{E} = -\frac{\partial \mathbf{A}'}{\partial t} \qquad (1.6)$$

---

[4] As a simple example consider Ohm's law for a long cylinder with non-homogeneous resistivity ($\mathbf{E} = \rho(x)\mathbf{J}$), and a slowly varying current along the length, giving a uniform current density $\mathbf{J}(t)$. The induced charge density will be given by $\rho_q = \epsilon_0 \nabla \cdot \mathbf{E} = \epsilon_0 \nabla\rho \cdot \mathbf{J}(t)$. Thus, we obtain $\rho_q(x, t) = \epsilon_0 \rho'(x)J(t)$ in the approximation $\nabla \cdot \mathbf{J} \approx 0$.

In view of our previous example for the ohmic conductor, the answer seems simple. One should be sure that $\nabla \cdot \mathbf{E}$ gives the correct charge density, and this depends on the correct selection of $\mathbf{A}'$. Here, we recall that a large flexibility exists for such selection (gauge invariance), but one must still keep in mind that the physical problem imposes some restrictions. Thus, if one chooses a certain vector potential $\mathbf{A}$ that gives the correct magnetic field, still must be sure that the correct electric field arises. In other words, perhaps a term of the kind $\nabla \phi$ must be added. Let us visualise this with an example taken from the theory of superconductivity.

Example. According to [5], the electric field that appears in a type-II super-conductor with square cross section ($|x| \leqslant a/2$, $|y| \leqslant a/2$) subjected to a time-dependent uniform magnetic field $(0, 0, B_a(t))$ is given by[5]

$$\mathbf{E} = (y - x)\dot{B}_a \hat{\boldsymbol{j}}, \quad 0 < y < x \tag{1.7}$$

Let us suppose that we decide to use $\mathbf{A} = xB_a\hat{\boldsymbol{j}}$. This leads to the correct value of $\mathbf{B}$: $\nabla \times \mathbf{A} = B_a \hat{\mathbf{k}}$. Nevertheless, it is apparent that the physical electric field is not plainly obtained from $\mathbf{A}$

$$\mathbf{E} \neq \frac{\partial \mathbf{A}}{\partial t} \tag{1.8}$$

On the other hand, by choosing $\mathbf{A}' = (x - y)B_a/2\hat{\boldsymbol{j}}$ one correctly gets:

$$\begin{aligned} \nabla \times \mathbf{A}' &= B_a\hat{\mathbf{k}} \\ -\frac{\partial \mathbf{A}'}{\partial t} &= (y - x)\dot{B}_a\hat{\boldsymbol{j}} \end{aligned} \tag{1.9}$$

Thus, our thumb rule should be that $\mathbf{A}$ and ϕ may not be chosen independently. In practice, one may solve the magnetostatics of the problem by arbitrarily choosing $\mathbf{A}$, but transient regimes, induced $\mathbf{E}$ fields, etc need further consideration. As the reader may check, in the above example, the correct value of $\mathbf{E}$ for $\mathbf{A} = xB_a\hat{\boldsymbol{j}}$ appears only if one introduces $\phi = -B_a y^2/2$. Of course, this is just an illustrative example, and in a real problem one will not have the solution at hand. However, aided by symmetry considerations or other restrictions (knowledge of regions where $\mathbf{E}$ should be zero, etc) one may choose the correct function ϕ.

As it will be later shown (section 2.4.2), the gauge invariance is not only an issue that one should care about when posing equations for a problem, but also a consideration for exploring the underlying physics of a phenomenon.

Evaluation of energy. Focusing on the quasi-stationary regime of current transport, we will be concerned with the energy of the magnetic field. As customary, the basic relation is

$$U_{\mathrm{M}} = \frac{1}{2\mu_0} \int_{\mathbb{R}^3} B^2 dV \tag{1.10}$$

---

[5] Here, we will just use this equation as a *postulate* for illustration purposes. Notice that we only pay attention to one eighth of the superconductor.

Here, we want to emphasise that integration is over all space, as the magnetic field, though decreasing may be non zero even at arbitrary large distances from the sources.

In many instances (mainly when numerical evaluations are necessary) we will prefer the interpretation that energy concentrates in the sources and use

$$U_M = \frac{1}{2} \int_\Omega \mathbf{A} \cdot \mathbf{J} \, dV \qquad (1.11)$$

with Ω being the region occupied by the system of currents[6].

A further manipulation leads to the formula that gives the energy in terms of the currents themselves

$$U_M = \frac{\mu_0}{8\pi} \int_V \int_{V'} \frac{\mathbf{J}(\mathbf{r}) \cdot \mathbf{J}(\mathbf{r'})}{||\mathbf{r}_i - \mathbf{r}_j||} dV dV' \qquad (1.12)$$

from which one can obtain the *circuital* expression

$$U_M = \frac{1}{2} \sum_{ij} I_i M_{ij} I_j \qquad (1.13)$$

that builds upon the idea of describing the continuous system of currents as a superposition of elementary circuits, and uses Neumann's formula that defines the mutual inductance between two of such circuits (e.g. c_i and c_j)

$$M_{ij} = \frac{\mu_0}{4\pi} \oint_{c_i} \oint_{c_j} \frac{d\boldsymbol{\ell}_i \cdot d\boldsymbol{\ell}_j}{||\mathbf{r}_i - \mathbf{r}_j||} \qquad (1.14)$$

Equations (1.10)–(1.13) are just equivalent forms (obtained one from the other through standard vector calculus manipulations) of the same physical quantity. Depending on the actual application we will choose one or another. In particular, the interpretation in equation (1.13) will be extremely useful for our purposes. If one discretises the electromagnetic problem in terms of a system of unknown current elements, the energy is a quadratic function in the unknowns I_i and may be effectively handled by numerical means. Thus, many situations will be solved by straightforward matrix multiplication.

Two eventual comments are due before concluding this paragraph:

- If one would review the arguments that lead to equation (1.10), textbooks show that such energy comes from accelerating charges against electromotive forces. This accounts just for work done against electromagnetic induction. No consideration is done about kinetic energy of the charge carriers in the material. However, as it will be shown in section 2.2, this will be a relevant factor in superconducting materials.

---

[6] We note that for the stationary state condition $\nabla \cdot \mathbf{J} = 0$, the result of applying equation (1.11) is independent of the gauge selection for the vector potential $\mathbf{A} \rightarrow \mathbf{A} + \nabla\phi$.

- In addition to the reversible (recoverable) energy that is loaded in the system by creating the system of currents (and related magnetic field) one has to consider that some energy may be irreversibly lost (in fact, transferred to the underlying thermodynamic system in the form of heat). Such losses may be calculated in terms of the electric field law for the actual material. By now, we just recall that in the case of a conventional metallic coil $\mathbf{E} = \rho \mathbf{J}$. Recall that, in this case, losses are given by

$$W_{\text{LOSS}} = \frac{1}{2} \int \int \rho J^2 dV \, dt \qquad (1.15)$$

along the process of 'charging' the coil $J(t)$, and also in the subsequent stationary state with $J = $ constant. In section 1.2 this point will be treated with detail from a rather general point of view, so as to apply it to different conduction laws.

1.1.2 The field H and the magnetisation M

In the theory of magnetism, the distinction between $\mathbf{B}$ and $\mathbf{H}$ is crucial, and related confusion not unfrequent. Recall that $\mathbf{H}$ appears as the field whose **curl** is determined by the free current densities, i.e. those that are not bound to molecular magnetic moments[7]

$$\nabla \times \mathbf{H} \equiv \nabla \times \left(\frac{\mathbf{B}}{\mu_0} - \mathbf{M} \right) = \mathbf{J}_{\text{FREE}} \qquad (1.16)$$

As recalled by Griffiths [2], $\mathbf{H}$ must be considered a useful auxiliary field, being $\mathbf{B}$ the fundamental quantity. In fact, this author argues that primarily is the fact that bound currents are as physical as free currents, and theoretically, one could just use

$$\nabla \times \mathbf{B} = \mu_0 (\mathbf{J}_{\text{FREE}} + \mathbf{J}_{\text{BOUND}}) \qquad (1.17)$$

However, as for the general case, the material laws which are behind the determination of $\mathbf{J}_{\text{BOUND}}$ are by no means simple, the introduction of $\mathbf{H}$ is so useful.

Aiming at a clear distinction between the two fields, many authors prefer to use *magnetic flux density* for $\mathbf{B}$ and either *magnetic field intensity* or *magnetic field strength* for $\mathbf{H}$. As it will be seen below, in our case, the use of $\mathbf{B}$ would be sufficient, but in many cases $\mathbf{H}$ will be retained so as to accommodate to the literature. A traditional reason for the introduction of $\mathbf{H}$ is that in the laboratory one controls the free current through the coils, which in the end creates the 'experimental' field applied to the samples. In fact, one could use either $\mathbf{H}_a$ or $\mathbf{B}_a = \mu_0 \mathbf{H}_a$ when talking about this quantity.

As concerns the contribution of the materials under investigation to $\nabla \cdot \mathbf{H}$ or what is equivalent to the bound (magnetisation) currents, things may be quite more

---

[7] In particular the main difficulty relates to identifying the source of divergence for $\mathbf{H}$ that does depend on the existence of bound (magnetisation) currents, in fact: $\nabla \cdot \mathbf{H} = -\nabla \cdot \mathbf{M}$.

involved. Being focused on a specific set of problems in macroscopic superconductivity, and aiming at the highest simplicity as possible for our purposes, we proceed as follows:

Magnetic materials. We will either work with permanent magnets or soft paramagnets.

Permanent magnets (chapters 8 and 13) will be described by a specified value of magnetisation (constant and homogeneous) $\mathbf{M}_0$ and the related magnetic field calculated from

$$\mathbf{B}(\mathbf{r}) = \frac{\mu_0}{4\pi} \int_{S'} \frac{\mathbf{K}_M(\mathbf{r}') \times (\mathbf{r} - \mathbf{r}')}{\|\mathbf{r} - \mathbf{r}'\|^3} \, dS' \tag{1.18}$$

with $\mathbf{K}_M$ being the magnetisation surface current distribution, obtained as customary from

$$\mathbf{K}_M = \mathbf{M}_0 \times \hat{n} \tag{1.19}$$

Soft paramagnets (chapter 14) will be described in terms of their relative permeability as follows. Their surface currents (*a priori* unknown) may be determined through the equation for the boundary condition at the surface[8]

$$H^+_{\text{TANGENTIAL}} = H^-_{\text{TANGENTIAL}} \Rightarrow \frac{B^+_{\text{TANGENTIAL}}}{\mu^+} = \frac{B^-_{\text{TANGENTIAL}}}{\mu^-} \tag{1.20}$$

This condition, combined with equation (1.18) will be used to determine $\mathbf{K}_M$, and eventually, with the magnetisation currents known, $\mathbf{B}(\mathbf{r})$ at any point of space using equation (1.18) again.

Superconducting materials. In general, distinction between $\mathbf{B}$ and $\mathbf{H}$ may be rather complicated[9]. As it will be explained in chapter 2, superconductors may exhibit the so-called mixed state in which microscopic loops of current (vortices) exist side-by-side with macroscopic free current density that flows along the whole sample. Then, distinction between $\mathbf{J}_{\text{FREE}}$ and $\mathbf{J}_{\text{BOUND}}$ is due and not simple in general. Non-trivial theoretical considerations are needed to build a complete theory [6]. Here, being concentrated on specific experimental conditions, one may simplify the treatment. In fact, we deal with the following situations:

- *Superconductors in the Meissner state*, which expel the magnetic flux by means of macroscopic superficial currents ($\mathbf{J}_{\text{BOUND}} = 0$). Then, the material contributes with a magnetic field given by either (1.3) or (1.18), depending on the thickness of the layer. In general, this will occur for 'weak' applied fields.

- *Superconductors in the critical state*: in this regime, although penetrated by vortices, the material supports average macroscopic currents whose contribution is much more relevant. This will occur for 'strong' magnetic fields, well

---

[8] Here, (+) and (−) indicate neighbouring points close within and close without the surface of separation between two media of respective permeabilities μ^+ and μ^- (one of them could be vacuum and then $\mu = \mu_0$).

[9] The reader is addressed to chapters 2 and 3 for a basic theoretical approach to superconducting phenomena.

above a critical value named *lower critical field*, but not so intense so as to break superconductivity (see equation (12.44) and related discussion)

$$H_a \gg H_{c_1} \quad \Rightarrow \quad J_{\text{FREE}} \gg J_{\text{BOUND}} \quad \Rightarrow \quad \mathbf{B} \approx \mu_0 \mathbf{H} \qquad (1.21)$$

- *Investigation of single vortices.* Thermodynamic ensembles of current vortices of microscopic size, the counterpart of molecular magnetic moments, have to be studied recalling the above concepts (**B**, **H**). However, for the case of having a single vortex as the object of interest, the situation simplifies and one may plainly speak about the magnetic field **B**. In some cases, for explicit indication of the microscopic nature, authors use lower case **b**.

Magnetisation. To end this section, we introduce the concept that straightforwardly connects to macroscopic experiments, i.e. the sample's magnetisation. Having stated that our description of the material relies on the knowledge of the underlying current density distribution, we will use the following relation for calculating the sample's full magnetic moment:

$$\boldsymbol{\mu} = \frac{1}{2} \int_\Omega (\mathbf{r} \times \mathbf{J}) \, dV \qquad (1.22)$$

Then, what is understood by sample's magnetisation or volume magnetisation is just

$$\mathbf{M}_V \equiv \frac{\boldsymbol{\mu}}{\text{sample's volume}} \qquad (1.23)$$

a quantity that one may obtain by a number of methods in the laboratory (as briefly summarised in section 4.2).

We want to stress that $\mathbf{M}_V$ is a macroscopic quantity not be confused with the mesoscopic function $\mathbf{M}(\mathbf{r})$ defined by molecular magnetic moments. For the readers' sake, we comment that in many instances, the notation $\mathbf{M}$ with subindex $_V$ dropped is used in the literature as one may easily deduce from the context. In this book, we will keep it.

Summarising this section, the physical problems presented in this book will be stated in terms of the fields $\mathbf{B}(\mathbf{r})$, $\mathbf{E}(\mathbf{r})$ and $\mathbf{J}(\mathbf{r})$ and potentials $\mathbf{A}(\mathbf{r})$ and $\phi(\mathbf{r})$. Though not strictly necessary, we retain $\mathbf{H}_a(\mathbf{r})$ for the contribution of external sources (typically coils) as an extended use in the literature. This emphasises that one has control over this field. Concerning the magnetisation of the samples, we will use the macroscopic value $\mathbf{M}_V$, i.e. the averaged magnetic moment over the volume.

1.2 The MQS regime

As said above, the Maxwell equations, as stated in equation (1.2), must be completed with the material law for the electric field in order to get the full set of

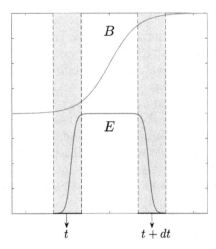

Figure 1.1. Schematics of the time dependence of the electromagnetic fields in the MQS approximation.

equations that determine our quantities of interest, i.e. the fields $\mathbf{B}(\mathbf{r})$, $\mathbf{J}(\mathbf{r})$, $\mathbf{E}(\mathbf{r})$. The purpose of this section is to offer a general approach for this problem. To start with, we introduce a non-conventional interpretation of the linear (ohmic) case, i.e. $\mathbf{E} = \rho\mathbf{J}$, that will be the basis for generalisation.

This section is partly based on the material exposed in the book by Orlando and Delin [4]. These authors give a thorough description of the so-called magneto-quasistatics (MQS) in a linear material. Specifically, they provide several examples of the solution of the low-frequency Maxwell equations by using Ohm's law.

As sketched in figure 1.1, the MQS approximation neglects the duration of the establishment of electric fields when some process occurs. In the figure, the relevant time interval dt is assumed to be noticeably bigger than the typical time for the appearance (disappearance) of $\mathbf{E}$, that remains basically constant along dt. Possible charge accumulation would take place in the negligible time intervals at the beginning/end of the interval.

1.2.1 Magnetic diffusion in conductors

Macroscopic superconductivity is tightly related to the phenomenon of magnetic field penetration under various conditions. As the analysis of practical situations will require familiarity with some physical concepts and mathematical tools that could be scarcely familiar to the reader, we devote this section to introduce such issues, accompanied by the widespread knowledge about normal metals. A generalised treatment valid for arbitrary conduction laws will be introduced here for the case of ohmic media. This will serve as a basis for application to superconductors in forthcoming chapters.

According to the MQS approximation introduced in section 1.1, at the macroscopic level, the dynamics of magnetic fields that penetrate in conducting media may be resolved through the Maxwell equations in (1.2). In practice, MQS means that charge accumulation may exist, but that the related processes may be skipped over for occurring quasi-instantaneously in the time scale of interest. Then, transient electric fields appear as long as magnetic flux variations exist and may be obtained by solving the equations under suitable boundary conditions. Eventually, if required, accumulated charges may be evaluated through the operation[10] $\nabla \cdot \mathbf{E}$ (see [5]). Concerning the time scale for the penetration of magnetic fields, it will be established by the reconfiguration of induced currents, which may be modelled through some macroscopic conductivity law for the material. Thus, the Maxwell equations must be supplemented with some law describing the conductivity (or superconductivity) of the material. As is customary, we will adopt the phenomenological point of view that some meaningful dependence $\mathbf{E}(\mathbf{J})$ is at hand. Mathematically, provided that suitable boundary conditions are known, this will lead to a well defined problem in terms of the fields.

Let us show, for instance, how one obtains an explicit differential statement for the problem. Thus, if one uses the material law in terms of some given resistivity

$$\mathbf{E} = \rho(\mathbf{J}) \, \mathbf{J}, \tag{1.24}$$

which emphasises the possible non-linearities that must be considered for super-conducting media, one ends up with the diffusion equation

$$\left(\mu_0 \frac{\partial}{\partial t} - \rho(\mathbf{J}) \nabla^2 \right) \mathbf{B} = (\nabla \times \mathbf{B}) \times \nabla \rho(\mathbf{J}) \tag{1.25}$$

As reported by several authors [7, 8], the difficulties of solving this statement by analytical methods are apparent, especially if one considers an arbitrary dependence of the $\mathbf{E}(\mathbf{J})$ law. The well known exception is the linear case, i.e. ρ = constant that leads to

$$\mu_0 \frac{\partial \mathbf{B}}{\partial t} - \rho \nabla^2 \mathbf{B} = 0 \tag{1.26}$$

for the diffusion of magnetic fields in normal metals. For this reason, alternative formulations will be explored below.

A prevalent technique in many physical theories is the variational interpretation that provides a formulation equivalent to the differential equations. Recall the celebrated case of classical mechanics: Hamilton's 'least action' principle allows to determine the physical trajectory chosen by a moving particle (among infinite possibilities) in the evolution from some initial state to a given final state [9][11]. Here, we will show that such an idea may be applied for obtaining a model equivalent to equation (1.25) that, in passing, will provide a number of advantages, allowing to include a wide class of physical problems represented by a variety of $\rho(\mathbf{J})$ laws. However, the situation is a bit more involved than for the most popular mechanical problems for which the reader may recall Hamilton's principle and Lagrange equations. On one side, here we are dealing with a field theory, which involves

---

[10] It is to be noticed that equations (1.2) only determine the rotational part of the electric field. Its divergence, if non zero, will be implicit in the material conductivity law.

[11] For those who are not familiar to such methods, a brief introduction, dedicated to our basic needs in this book, is compiled in section 5.1.

Lagrangian densities. For the electromagnetic case in the MQS limit, that neglects the contribution of energy storage in the form of electrostatic fields, one has the following action integral

$$S_{em} \approx \int \mathcal{L}_{mqs} \, d^4\mathbf{x}, \quad \mathcal{L}_{mqs} \equiv -\frac{1}{2\mu_0} B^2 + \mathbf{J} \cdot \mathbf{A}. \qquad (1.27)$$

On the other side, strictly speaking, the minimisation of S_{em} in equation (1.27) gives the equations for the magnetic field in vacuum in the presence of a prescribed source of current $\mathbf{J}$. The introduction of a material law describing the interactions within the conductor must be elaborated. For this purpose, the main difficulty relates to the fact that here we must deal with a non-conservative system, i.e. one for which some fraction of the internal electromagnetic energy is 'dissipated' for being supplied to some other sector.

For the incorporation of this physical phenomenon one needs to consider some introductory concepts of irreversible thermodynamics that are dealt with below[12].

1.2.2 Magnetic diffusion: thermodynamic background

One of the main fields of interest of non-equilibrium thermodynamics, the study of the so-called steady states, will guide us through the analysis of electric conduction. In fact, being interested in quasi-steady systems, the extension of the 'steady state' techniques to quasi-steady transient processes (magnetic diffusion) will be straight-forward. Specifically, along the lines established above, the evolutionary processes will be described by a concatenation of time intervals, each of them having an effectively constant (steady) electric field and current density (figure 1.1).

As an example of the kind of arguments that will be used below, we recall [11] that relying on the law of increasing entropy and also on Onsager reciprocal relations, one can show that in normal conductors, a physically acceptable conductivity tensor must be positive definite and symmetric. This generalises the one dimensional version of Ohm's law. Certainly such a kind of conclusion could never be reached by pure electrical or mechanical principles.

To begin with, we will recall some definitions and introduce the basic principles based on the physics of normal conductors. Application to generalised conduction problems will be apparent. First, as a local description is required for our purposes, we introduce a mesoscopically averaged entropy function (per unit volume) $\mathscr{S}$. It is assumed that microscopical fluctuations are smeared, and $\mathscr{S}$ may be considered a smooth function, depending on the local current density, i.e. $\mathscr{S}[\mathbf{J}(\mathbf{r})]$. When the conductor is free from external excitations it will settle at the stable equilibrium point, that according to the second law of thermodynamics is given by

$$\mathscr{S}_{eq} = \mathscr{S}[J = 0] = \max \{\mathscr{S}[J]\}$$

---

[12] A more detailed exposition, developed by C López may be found in [10].

Owing to the statistical nature of the problem (huge number of microscopical states for a single macroscopical state), fluctuations around $\mathscr{S}_{eq}$ will become negligible. Now, if the system is driven out of equilibrium by some external agent (but not very far from it) one may assume that internal mechanisms will operate causing a restoring force that moves the system back to equilibrium as soon as the action stops. This will produce an amount of heat that may be quantified through the entropy increase that takes place

$$dQ = Td\mathscr{S} \tag{1.28}$$

Our hypothesis will be that in the event of a maintained external action, the same internal thermodynamic forces will continuously act so as to produce entropy as given by the above expression. Then, one may define a rate of heat/entropy production through a dissipation function defined by $Td\mathscr{S} = 2\mathcal{P}dt$.[13] As isothermal conditions will be assumed, we will use the relation

$$\dot{\mathscr{S}} = 2\mathcal{P}/T \tag{1.29}$$

Now, according to the fundamentals of irreversible thermodynamics [12] one may apply the 'principle of minimum entropy production'.

> *The steady state of a system is that state in which the rate of entropy production has the minimum value consistent with the external constraints.*

Notice that, according to this statement, the equilibrium state naturally occurs when there are no constraints on the system, because as a consequence of the second law, it reaches the maximum entropy, and thus the absolute minimum of entropy production, i.e. zero.

Let us now see how the above principle applies to the case of electric conduction. Apparently, close to equilibrium, the dissipation function may be 'Taylor expanded' in the form

$$\mathcal{P} \simeq \frac{1}{2}\sum_{ij} J_i\, \mathrm{R}^{ij} J_j \tag{1.30}$$

with R^{ij} being the components of a symmetric positive definite tensor. Then, as desired, one gets the equilibrium condition $\mathbf{J} = 0$. From this equation, one may obtain the thermodynamic field drives the system towards equilibrium (towards the minimum value of $\mathcal{P}$ in terms of the variable $\mathbf{J}$)

$$\mathbf{E}_{ther} = -\nabla_J \mathcal{P} = -\mathrm{R}\,\mathbf{J}, \tag{1.31}$$

---

[13] The factor 2 will be clarified later on.

and considering that the steady state occurs by the balance between this field and the external action, the required field that enables this state is

$$\mathbf{E} = -\mathbf{E}_{\text{ther}} = \mathsf{R}\,\mathbf{J},\qquad(1.32)$$

i.e. one has a reinterpretation of Ohm's law for steady state conduction in terms of non-equilibrium thermodynamics.

Generalised conduction problems, as those that will be studied in chapter 8 for superconducting materials will be treated in terms of physically meaningful expressions of $\mathcal{P}[\mathbf{J}]$ alternative to equation (1.30).

Summarising, to this point we have issued minimisation principles (equations (1.27) and (1.30)) that may be 'separately' used to describe

- the magnetic field equations in free space and under controlled sources[14]
- the steady state of conducting media including dissipation of energy.

In the next section, we show that keeping the ansatz of slow variations, one may use a 'variational-type'[15] model to describe the quasi-steady evolution of the electromagnetic system with dissipation.

1.2.3 A simple mechanical analogue

In order to clarify the physical nature of the problem, we will first invite the reader to analyse a simple mechanical analogue, consisting of a single particle under the simultaneous action of a conservative force and a drag force.

Let us first recall the simplest 1D problem: a single particle moves subject to a field with associated potential energy U. The action integral corresponding to Hamilton's principle is [9] (see also section 5.1 in this book)

$$\mathsf{S} \equiv \int L(x,\,v)\,dt = \int \left[\frac{mv^2}{2} - U(x) \right] dt \qquad (1.33)$$

Then, the Euler–Lagrange equations become

$$\text{Min } \mathsf{S} \quad \Rightarrow \quad \frac{d}{dt}\frac{\partial L}{\partial v} - \frac{\partial L}{\partial x} = 0 \quad \Leftrightarrow \quad m\frac{dv}{dt} = -\frac{\partial U}{\partial x} \equiv F_{\text{cons}} \qquad (1.34)$$

as it is well known.

Consider now that a viscous drag $F_{\text{drag}} = -hv$ acts on the particle (see figure 1.2). Let us show that, under appropriate conditions, a minimum principle leading to the sound equations of motion can still be formulated in terms of a modified Lagrangian and action integral

---

[14] As it will be shown in section 3.2, the concept of 'free space' may be augmented to problems in which dissipation does not occur, for instance within a superconductor in the Meissner state.

[15] Quotations will be used because here the term *variational* will be used in a more general sense than in the usual context of classical mechanics. Strictly speaking, one should speak about a quasi-variational statement from the mathematical point of view [13].

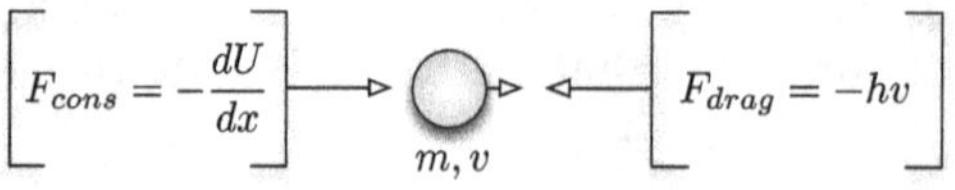

Figure 1.2. A single particle under the action of a conservative force and a viscous drag force.

$$\hat{S} \equiv \int \hat{L}(x, v, t)\, dt \quad \text{with} \quad \hat{L} = L + \mathcal{P}t \equiv L + \frac{1}{2}hv^2\, t \qquad (1.35)$$

$$\text{Then: Min } \hat{S} \;\Rightarrow\; \frac{d}{dt}\frac{\partial \hat{L}}{\partial v} - \frac{\partial \hat{L}}{\partial x} = 0 \;\Leftrightarrow\; m\frac{dv}{dt} \simeq -\frac{\partial U}{\partial x} - hv = F_{\text{cons}} + F_{\text{drag}}$$

We stress that, in deriving the above equations one has to neglect variations of the viscous force F_{drag} within the interval of time considered. The reader may check that the approximation will be valid if minimisation is applied 'iteratively' with intervals of duration much less than the characteristic time $\tau \equiv h/m$. This relates to the so-called *adiabatic hypothesis* recurrently used in other physical disciplines:

> *If energy, though not conserved, varies slowly according to some parameter, then one is allowed to assume a kind of isolated system within small enough intervals of the temporal evolution.*

A final calculation will give us some further physical insight. Let us use the canonical expression for calculating the energy of the particle in terms of the modified Lagrangian. Upon neglecting the term $\Delta F_{\text{drag}}\Delta x$ the energy within a given interval is

$$\mathcal{E} = v\frac{\partial \hat{L}}{\partial v} - \hat{L} \simeq \frac{mv^2}{2} + \frac{1}{2}F_{\text{drag}}v\Delta t, \qquad (1.36)$$

that is to say, within each interval one may visualise the particle as an isolated system which stores an average energy increased by one half of the full loss $F_{\text{drag}}v\Delta t$.

1.2.4 MQS quasi-steady states: thermodynamic background

As shown in [14] the quasi-steady evolution of an electromagnetic field interacting with a conducting medium may be described along the above lines. Skipping some technical details, which the interested reader may find in that reference, the basic ideas are as follows. First, consistently with the *adiabatic hypothesis* one performs time integration in the 4-integral of equation (1.27) so as to obtain a purely spatial statement. Physically, this means to divide time into intervals and consider the quantities of interest in the sense of averages. Then, dissipation is included through the function $\mathcal{P}(\mathbf{J})$ defined in section 1.2.3, ending up with the modified action integral valid for each 'time layer'

$$\mathcal{F}[\mathbf{B}_{n+1}(\mathbf{r})] = \frac{1}{2\mu_0} \int_{\mathbb{R}^3} \left\| \mathbf{B}_{n+1} - \mathbf{B}_n \right\|^2 dV + \int_\Omega \Delta t\, \mathcal{P} dV \qquad (1.37)$$

with Ω being the region occupied by the conductor and $\mathbf{B}_{n+1}$ the unknown magnetic field. Notice that $\mathbf{B}_n$ means the magnetic field distribution at a given time instant given by t_n and $\Delta t \equiv t_{n+1} - t_n$.

To start with, let us apply this formulation to the case of a normal conductor. One can show that the Euler–Lagrange equations lead to the desired diffusion equation (equation (1.26)). Thus, if one assumes a diagonal resistivity matrix, i.e. $\Omega^{ij} = \rho\, \delta^{ij}$, replaces $\mathcal{P}$ by $\rho J^2/2$ (equation (1.30)) and uses Ampère's law, it follows

$$\text{Min}\ \ \mathcal{F}\,[\mathbf{B}_{n+1}(\mathbf{r})]$$
$$\Downarrow$$
$$\frac{\partial \hat{\mathcal{L}}}{\partial \mathbf{B}_{n+1}} = \sum_j \frac{\partial}{\partial x_j} \frac{\partial \hat{\mathcal{L}}}{\partial(\partial \mathbf{B}_{n+1}/\partial x_j)} \qquad (1.38)$$
$$\Downarrow$$
$$\mu_0 \frac{\mathbf{B}_{n+1} - \mathbf{B}_n}{\Delta t} = -\rho \nabla \times \nabla \times \mathbf{B}_{n+1} = \rho \nabla^2 \mathbf{B}_{n+1}$$

i.e. the time discretized form of equation (1.26).

Outstandingly, the interest of equation (1.37) is that it gives way to the possibility of applying direct numerical minimisation methods[16], as an alternative to the differential equation statements. In particular, the direct variational statement allows to deal with non-simple forms of $\mathcal{P}(\mathbf{J})$ (sometimes ill-posed from the mathematical point of view) as those related to the investigation of type-II superconductors.

1.2.5 Physical interpretation

Equation (1.37) may be interpreted in similar fashion to the mechanical analogue introduced before. The evolution of the system is realised by the balance between an inertial term, and a dissipation term. By separation, the former would lead to the static situation $\mathbf{B}_{n+1} = \mathbf{B}_n$, while the second would give rise to the steady state of conduction (as shown in exercise R1-3).

---

[16] In many applications, the functional in equation (1.37) will be more conveniently reexpressed in terms of the current densities. As shown in chapter 8 this will allow to restrict the integration to the volume of the sample, a fact that obviously eases numerical calculations.

1.3 Review exercises

Review exercises R1-1

It is common practice in textbooks to show that the expression

$$A = \frac{1}{2}B \times r \tag{1.39}$$

gives the vector potential for a uniform electromagnetic field. Also known is that one may apply gauge transformations and that $A' = A + \nabla f$ may also be used with f some scalar function. Show that for the case $B = B_0 \hat{j}$ one may either use $A = (B_0/2)(z, 0, -x)$ or $A' = B_0(0, 0, -x)$ and find the corresponding function f.

Hint: check $\nabla \times A$, $\nabla \times A'$ and inspect $A - A'$.

Review exercises R1-2

It is also common practice in EM books to show that the magnetic field of the uniformly magnetised sphere is dipolar outside, uniform inside. Review the structure of the vector fields B, H, M. It may be checked numerically by using the software for figure 9.2. See what happens for an incomplete sphere (half-sphere for instance).

Hint: in the scripts provided, replace the position dependent surface current density for the superconductor by $K(\theta) = M_0 \sin(\theta)$ with θ being the polar angle.

Review exercises R1-3

Show that, in a normal conducting material the distribution of current's density is such that entropy production is minimum in steady state conditions.

Hint: use (1.30) and minimise the production of entropy with the restriction $\nabla \cdot J = 0$ [11]. Recall that restrictions are introduced by means of Lagrange multipliers.

Review exercises R1-4

Transform the functional $\mathcal{F}[\mathbf{B}_{n+1}(\mathbf{r})]$ in equation (1.37) by 'changing variables' to $\mathcal{F}[\mathbf{J}_{n+1}(\mathbf{r})]$. Show that it takes the form

$$\mathcal{F}[\mathbf{J}_{n+1}(\mathbf{r})] = \frac{\mu_0}{4\pi} \int \int_{\Omega} \left[\frac{1}{2} \frac{\mathbf{J}_{n+1}(\mathbf{r}) \cdot \mathbf{J}_{n+1}(\mathbf{r}') - 2\mathbf{J}_n(\mathbf{r}) \cdot \mathbf{J}_{n+1}(\mathbf{r}')}{\|\mathbf{r} - \mathbf{r}'\|} \right] dVdV'$$
$$+ \int_{\Omega} \Delta\mathbf{A}_a(\mathbf{r}) \cdot \mathbf{J}_{n+1}(\mathbf{r}) \, dV + \Delta t \int_{\Omega} \mathcal{P}[\mathbf{J}_{n+1}(\mathbf{r})]dV$$

with $\mathbf{A}_a$ being the vector potential contributed by the external sources (*applied*).

Hint: use Ampère's law, the properties of differential vector operators and the integral theorems.

References

[1] Jackson J D 1975 *Classical Electrodynamics* 2nd edn (New York: Wiley)

[2] Griffiths D J 2013 *Introduction to Electrodynamics* 4th edn (Boston, MA: Pearson)

[3] London H 1936 *Proc. Roy. Soc.* A **155** 102–10

[4] Orlando T P and Delin K A 1991 *Foundations of Applied Superconductivity* (Reading, MA: Addison-Wesley)

[5] Brandt E H 1995 Electric field in superconductors with rectangular cross section *Phys. Rev.* B **52** 15442–57

[6] Tinkham M 1996 *Introduction to Superconductivity International Series in Pure and Applied Physics* (New York: McGraw-Hill)

[7] Cha Y S 2000 Magnetic diffusion in high-T_c superconductors *Physica* C **330** 1–8

[8] Surdacki P 2003 Modeling of the magnetic field diffusion in the high-T_c superconducting tube for fault current limitation *Physica* C **387** 234–8

[9] Goldstein H 1980 *Classical Mechanics* (Reading, MA: Addison-Wesley)

[10] Badía-Majós A and López C 2012 Electromagnetics close beyond the critical state: thermodynamic prospect *Supercond. Sci. Technol.* **25** 104004

[11] Landau L D and Lifshitz E M 1984 *Electrodynamics of Continuous Media* 2nd edn (Oxford: Butterworth-Heinemann) vol 8 of (Course of Theoretical Physics)

[12] Prigogine I 1967 *Introduction to Thermodynamics of Irreversible Processes* (New York: Interscience)

[13] Berdichevski V L 2009 Interaction of Mechanics and Mathematics *Variational Principles of Continuum Mechanics* (Heidelberg: Springer)

[14] Badía-Majós A, Cariñena J F and López C 2006 Geometric treatment of electromagnetic phenomena in conducting materials: variational principles *J. Phys. A: Math. Gen.* **39** 14699–726

IOP Publishing

Macroscopic Superconducting Phenomena
An interactive guide
Antonio Badía-Majós

Chapter 2

Basic superconductivity: observations and theories

This chapter introduces the basic physical properties of superconducting materials, together with the phenomenological theories that allow to issue mathematical models for their description. The statements will be made in terms of physical parameters (characteristic lengths) whose ultimate microscopic nature is not put on focus. In brief, we will review the amazing electromagnetic properties, i.e. zero resistance and magnetic flux expulsion, as well as the stunning manifestation of the quantum nature of the phenomenon: flux quantisation in particular. An overview of the London equations at a purely classical level and their 'transition' to the quantum statement, i.e. the Ginzburg–Landau theory, will be presented. With this at hand, and following the customary programme, we will recall the classification of super-conductors in type-I and type-II materials, with some emphasis on the latter for further applications. In particular, the physics of flux vortices will be reviewed. Eventually, a list of typical materials and their specific properties will be presented.

2.1 Basic phenomenology

Superconductivity was first reported in 1911 by H Kamerlingh Onnes [1–3], who observed the sudden disappearance of electrical resistance in a filament of mercury in a liquid helium bath ($T_c \approx 4.2\ K$). As remarked by Kamerlingh Onnes and emphasised in figure 2.1, the outstanding fact is the steep jump from finite resistance to an indetectable value which, in fact, is an indication of a transition to a new state: the material becomes a *perfect conductor*, when the temperature goes below a certain value, the so-called *critical temperature* T_c.

An intense activity along the subsequent years soon gave place to new discoveries. In particular, it was clarified that the transition to the zero resistance state was influenced by the application of magnetic fields ($\mu_0 H_a$). Along this line, in another pathbreaking experiment, Meissner and Ochsenfeld [4] showed that superconductivity

doi:10.1088/978-0-7503-2711-4ch2

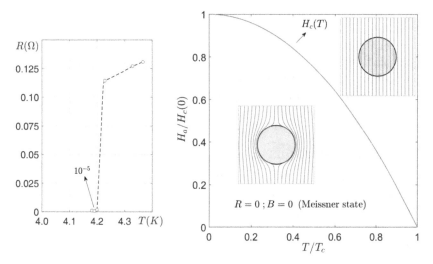

Figure 2.1. Left: transition to zero resistance in Hg (reproduced from [3], with permission of Springer). Right: sketch of the expulsion of magnetic fields by a superconductor and the transition line to the non-superconducting state.

is something more than perfect conductivity. They reported that the transition is accompanied by the *expulsion of magnetic flux*, i.e. within the superconductor one has $\mathbf{B} = 0$. In other words, the material behaves as a perfect diamagnet.

In figure 2.1 we illustrate the flux expulsion property in a superconductor under a uniform magnetic field. Another relevant information that one may extract from this figure is the definition of the so-called *critical field* H_c. In general, the suppression of superconductivity is realised by a combination of temperature and magnetic field. Thus, even at $T = 0$, the application of fields $H_a \geqslant H_c(0)$ produces the breakdown of the superconducting state[1]. As a matter of fact, the transition is defined by a phase diagram as shown in the plot, where we have used the empirical parabolic relation

$$H_c(T) = H_c(0)\left[1 - \left(\frac{T}{T_c}\right)^2\right] \qquad (2.1)$$

Such dependence agrees with the prediction of the phenomenological two-fluid model issued by Gorter and Casimir in 1934 [5]. This equation is of much conceptual interest because it relates straightforwardly to the microscopic nature of the phenomenon. Just recall that one may interpret the condensation of the charge carriers to the superconducting state just through the reduction of the material's free energy density by the amount $\mu_0 H_c^2/2$ as related to the expulsion of magnetic flux.

In the forthcoming sections we summarise the theoretical rationale that emerged from the above facts and that was developed through the decades in conjunction

---

[1] Strictly speaking, the intrinsic material characteristic H_c may strongly differ from the critical experimental value H_a (applied magnetic field) owing to finite size effects. This point is considered in section 3.1 and extensively in chapter 9.

with new experiments and discoveries. As in the rest of the book, we will keep at the phenomenological level and restricted to the main facts concerning our scope. Many of the forementioned discoveries relate to the search of superconducting materials with higher and higher critical temperatures and fields (which were really low at the time of the above referred experiments). At the end of the chapter (section 2.8) we will include a classification of some representative materials and their characteristic parameters.

2.2 The London equations

Perfect conductivity and perfect diamagnetism are well described at the level of macroscopic electromagnetism by the so-called *first* and *second London equations*. These phenomenological expressions were proposed by the London brothers in 1935 [6] and are usually displayed as follows. The first London equation (perfect conductivity) reads:

$$\mathbf{E} = \frac{\partial}{\partial t}(\Lambda_L \mathbf{J}) \tag{2.2}$$

which obviously means: $\mathbf{J}$ is time independent $\Rightarrow \mathbf{E} = 0$, i.e. constant supercurrents may exist in the material.

The second London equation (perfect diamagnetism) reads:

$$\nabla \times \Lambda_L \mathbf{J} = -\mathbf{B} \tag{2.3}$$

Flux expulsion is not that obvious from this equation. However, it becomes clearer upon using Ampère's law and the definition[2]

$$\Lambda_L \equiv \mu_0 \lambda_L^2 \tag{2.4}$$

One gets:

$$\lambda_L^2 \nabla \times \nabla \times \mathbf{B} = -\mathbf{B} \Rightarrow \lambda_L^2 \nabla^2 \mathbf{B} = \mathbf{B} \tag{2.5}$$

At first glance, this equation admits exponentially decreasing solutions of the kind $B \simeq \exp(-x/\lambda_L)$ in an ideal 1D problem. In chapter 3 we will explicitly solve it for a non-trivial, though still analytic statement that allows to quantify the flux expulsion phenomenon in a real experiment. Numerical procedures for complex geometries will be provided in chapters 9 and 13.

On the other hand, one may obtain an interesting physical interpretation of the above expressions by using a classical picture. Thus, equation (2.2) was defined in [6] as an 'accelerating' equation and may be straightforwardly related to the transport of current in a medium for which thermodynamic fields that drive a normal conductor towards equilibrium (equation (1.31) in section 2.2) are absent. Then, charge carriers will be accelerated as long as an electric field is present. We notice

---

[2] Recall that λ_L has the units of a length. It is the so-called London's penetration depth, whose significance will be analysed below.

that equation (2.2) just follows by applying the relation $\mathbf{J} = nq\mathbf{v}$ and Newton's second law

$$m\frac{d\mathbf{v}}{dt} = q\mathbf{E} \quad \Rightarrow \quad \Lambda_{\text{L}}\frac{d\mathbf{J}}{dt} = \mathbf{E} \tag{2.6}$$

if one defines

$$\Lambda_{\text{L}} \equiv \frac{m}{nq^2} \tag{2.7}$$

with, m, q, n the effective mass, charge and density of whatever represents the charge carriers in the material.

Keeping the analysis at the classical physics level, one may get a deeper insight by using conventional relations in the electromagnetic theory. To start with, we notice that according to equation (2.2) one has[3]

$$\mathbf{E} \cdot \mathbf{J} = \Lambda_{\text{L}}\mathbf{J} \cdot \frac{d\mathbf{J}}{dt} = \frac{d}{dt}\left(\frac{\mu_0\lambda_{\text{L}}^2 J^2}{2}\right) = \frac{d}{dt}\left(\frac{nmv^2}{2}\right) \tag{2.8}$$

In other words, by accelerating charges, the electric field supplies a kinetic energy that will be stored by the supercurrent.

A step further in the physical interpretation that will be of help to upgrade the theory follows by considering the full energy of the system formed by the super-conducting carriers and the magnetic field. Notice that one may write

$$U = \int_{\mathbb{R}^3} \frac{B^2}{2\mu_0}dV + \int_{V_{\text{S}}} \frac{\lambda_{\text{L}}^2}{2\mu_0}|\nabla \times \mathbf{B}|^2 dV \tag{2.9}$$

with the energy U including the magnetostatic field energy as well as the kinetic part described above.

Then, the equilibrium configuration of the system may be obtained by seeking the minimum of energy. This can be done by applying the functional analysis techniques called upon in section 1.2 and reviewed from the mathematical side in section 5.1. The configuration $\mathbf{B}(x, y, x)$ compatible with prescribed boundary conditions may be obtained from

$$\frac{\delta U}{\delta \mathbf{B}} = 0 \Rightarrow \frac{\partial u}{\partial B_{x_i}} = \sum_j \frac{\partial u}{\partial(\partial B_{x_i}/\partial x_j)} \tag{2.10}$$

where x_i means either x, y, z and u stands for the energy density.

Recall that the application of equation (2.10) leads to the flux expulsion equation

$$\mathbf{B} + \lambda_{\text{L}}^2 \nabla \times \nabla \times \mathbf{B} = 0 \tag{2.11}$$

---

[3] Steady or quasi-steady conditions are assumed, so that one can use total time derivatives.

This expression deserves several comments:

- Just by using Ampère's law ($\nabla \times \mathbf{B} = \mu_0 \mathbf{J}$) one straightforwardly goes back to (2.3). Thus, starting with the first London equation and recalling classical electromagnetic formulation one obtains

$$\mathbf{B} + \Lambda_{\mathrm{L}} \nabla \times \mathbf{J} = 0 \qquad (2.12)$$

  i.e. the second London equation.

- Equation (2.11) and the minimisation of (2.9) are equivalent statements of the Meissner effect, i.e. flux expulsion. λ_{L} clearly appears as the characteristic length scale that 'controls' the penetration of magnetic fields. For practical purposes, the variational statement (i.e. minimise $U(\mathbf{B})$) will be very useful. On the other hand, in both cases, one will have to care about the boundary conditions to be applied, i.e. $\mathbf{J}$ is confined to the volume of the superconductor and $\mathbf{B}$ has to include the information of the external sources. Both aspects will be considered in our proposed exercises.

- Although all the above is expressed in macroscopic variables, one may be expecting some underlying link to the microscopic nature of the phenomenon. In fact, the reader could well deplore the above discussion for being a somehow tricky mixture of field theory and classical mechanics. In particular, the role of the material is somehow blinded when one derives equation (2.11) by using the field $\mathbf{B}$ as a variable. On the other hand, by careful inspection of the several shortcuts taken above we will not only improve the range of application, but also guide us towards the microscopic mechanisms. In passing, this will give us the chance to unveil another fascinating property exhibited by superconductors: the *macroscopic manifestation of the quantum nature* of the phenomenon.

- Implicitly, Λ_{L} (equivalently λ_{L}) has been considered as a constant along this section; inhomogeneities within the sample and anisotropy have also been disregarded. The considerations necessary so as to include these properties will be introduced in the exercises of the chapter.

2.3 From classical to quantum

The above reviewed London equations are phenomenological expressions introduced as a good description of the experimental facts. As such, they ignore the microscopic nature of the phenomenon and do not have the capacity of establishing relations between different properties[4] or their nature. Nevertheless, as already noticed by F. London in the late 1940s, the careful inspection of these equations may pave the path to the underlying physical mechanisms, establish such relations and predict new phenomena. This section contains an overview of such analysis, that will also serve us for further applications.

Owing to the nature of superconductivity, quantum mechanics will be an essential element to understand and even compute macroscopic phenomena. Some essentials are to be called upon.

---

[4] To this point, zero resistance and flux expulsion are independent facts.

2.4 Wave functions: the Ginzburg–Landau theory

Since the early 1930s, the phenomenological theory introduced in the previous section has been of much use for the description of many aspects of the macroscopic manifestations of superconductivity. Nevertheless, it was immediately recognised that a microscopic theory would allow both to incorporate new aspects of the phenomenon as well as to better understand its nature. As emphasised by F London in his book [7], the construction of an electronic theory of superconductivity withstood many unsuccessful efforts along years. A work by this author published in 1948 [8] may be considered as a very noticeable step forward. Although still ignoring the microscopic origin of the interaction[5], he already envisioned its relation with the macroscopic manifestations that were well described by his phenomenological equations. Below, we will put forward a programme inspired by the basic ideas introduced by London, and that ends up with the second milestone in the history of superconductivity, i.e. the phenomenological Ginzburg–Landau (GL) equations. In brief, the GL equations follow naturally from the London equations when one describes the material in terms of a new field (given by a 'wave function' Ψ) that is conveniently coupled to the electromagnetic field. This will bring forward the quantum nature of the phenomenon.

2.4.1 Gauge invariance: the current density ω

Taking a closer look at equations (2.2) and (2.3) one may already advance some fingerprints of the microscopic nature. To start with, we recall that both equations may be obtained by using the following ansatz for the material law

$$\mathbf{J} = -\frac{\mathbf{A}}{\Lambda_{\mathrm{L}}} \Rightarrow \begin{cases} \dfrac{\partial}{\partial t}(\Lambda_{\mathrm{L}}\mathbf{J}) = -\dfrac{\partial \mathbf{A}}{\partial t} = \mathbf{E} + \nabla\phi \\[2mm] \nabla \times (\Lambda_{\mathrm{L}}\mathbf{J}) = -\nabla \times \mathbf{A} = -\mathbf{B} \end{cases} \tag{2.13}$$

Thus, regardless the electrostatic part of $\mathbf{E}$, both London equations may be unified in terms of the vector potential. As said by London, this could be interpreted by saying that the acceleration of charges ($\partial_t \mathbf{J}$) only depends on the part of $\mathbf{E}$ with subtracted scalar potential ($-(-\nabla\phi)$).

The above interpretation is attractive but a serious objection may be submitted. $\mathbf{J}$ is a physical quantity and thus should remain unchanged upon gauge transformations: $\mathbf{A} \to \mathbf{A}' = \mathbf{A} + \nabla\chi$. On the other hand, recalling charge conservation[6] in the MQS regime (magnetoquasistatic, see chapter 1) ensures the condition $\nabla \cdot \mathbf{J} = 0$. In view of this, one may propose to use the more general statement, i.e.

$$\mathbf{J} = -\frac{\mathbf{A}}{\Lambda_{\mathrm{L}}} + \omega \tag{2.14}$$

---

[5] This would still await for a decade, until the seminal contribution by Bardeen, Cooper and Schrieffer [9].
[6] Recall that, in electromagnetism, gauge invariance is in fact the counterpart of charge conservation (as related to Noether's theorem), and thus a requirement for any theory.

with $\boldsymbol{\omega}$ being proportional to the gradient of some function. This may solve the above requirements. For further convenience, let us define it by

$$\boldsymbol{\omega} \equiv \frac{\nabla \chi}{\Lambda_L} \tag{2.15}$$

Notice that, to this point, $\boldsymbol{\omega}$ is basically a mathematical artefact that one should include in the theory so as to adjust the possible gauge selection for the vector potential. Recall, for instance, that without this element one could not use a gauge selection other than $\nabla \cdot \mathbf{A} = 0$, which sounds unacceptable from the physical point of view.

Through basic dimensional arguments, one may advance that $\boldsymbol{\omega}$ could represent a certain current density related to the superconducting phenomenon, and χ a connected physical property. Some further analysis and accompanying experiments will reveal their origin. Thus, the improved superconducting model described by the updated expression (2.14) unifies zero resistance and flux expulsion and suggests to consider the reinterpretation of the electrical current density. In order to have further hints on the possible physical origin of this fact, we will introduce a new mathematical element, that will immediately lead to some plausible physical picture.

Figure 2.2 represents a region within a superconducting material that contains a singular point (central point). A closed integration path (originally square) around such point is defined and named $\mathcal{C}$. Here, by 'singular' point we mean some part of the material (no matter its size in fact) that for some reason is not superconducting. For instance, it could represent a void within the sample. What matters here is that the material law (equation (2.14) in our case) does not hold in such part. We are planning to analyse the consequences of measuring physical properties in such region. Recall that the relation between potentials and physical fields is established by path and surface integration.

Concerning the integration of $\mathbf{J}$ and $\mathbf{A}$ one may proceed without problems along $\mathcal{C}$, as continuity of the electromagnetic fields is expected. Nevertheless, as the role of $\boldsymbol{\omega}$ and χ is expected to connect to the properties of the material, one may suspect that

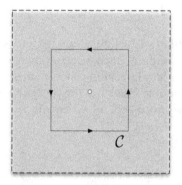
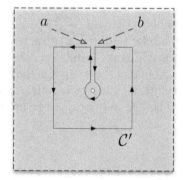

Figure 2.2. Integration path ($\mathcal{C}$) defined within the superconducting material with a central singularity (e.g. a void). To the right, integration around the singularity is avoided by the trajectory $\mathcal{C}'$ containing a tight circular roundabout. Points a and b are assumed to be at infinitesimal distance.

integration is affected by the presence of the void. In order to expedite quantitative evaluations we proceed by defining a deformed path, i.e. C' that includes a tiny turnabout around the void. Thus, one gets

$$\oint_C \mathbf{J} \cdot d\ell = -\frac{1}{\Lambda_L} \oint_C \mathbf{A} \cdot d\ell + \oint_C \boldsymbol{\omega} \cdot d\ell \qquad (2.16)$$

which straightforwardly transforms to

$$\oint_C \mathbf{J} \cdot d\ell + \frac{1}{\Lambda_L} \iint_S \mathbf{B} \cdot d\mathbf{s} = \oint_C \boldsymbol{\omega} \cdot d\ell \qquad (2.17)$$

where we have introduced the area enclosed by C and used Stokes' theorem for the magnetic flux[7].

The reader may wonder about the necessity (and utility) of this analysis. As said above, it will enable a physical interpretation. Thus, the importance of the void relates to the question: should not one expect that the role of the function ω (that depends on some material property) is to leave some fingerprint in the path integration? In other words, are we allowed to straightforwardly assume

$$\oint_C \boldsymbol{\omega} \cdot d\ell = \frac{1}{\Lambda_L} \oint_C \nabla\chi \cdot d\ell \stackrel{?}{=} 0 \qquad (2.18)$$

In order to check this, we use the auxiliary path C'. As C' only surrounds superconducting material, where ω should be well defined, one may write

$$\oint_{C'} \boldsymbol{\omega} \cdot d\ell = 0 \Rightarrow \frac{1}{\Lambda_L} \oint_C \nabla\chi \cdot d\ell = \frac{1}{\Lambda_L} \oint_o \nabla\chi \cdot d\ell \qquad (2.19)$$

where the integration along the periphery of the void ($\oint_o$) is not necessarily 0, as one may visualise in the following example. In fact, as we will see later there is a strong analogy between both cases.

Example
Let θ represent the polar angle in the XY plane. It is apparent that integration around the origin of coordinates gives

$$\oint_C \nabla\theta \cdot d\ell = \oint_o \nabla\theta \cdot d\ell = 2\pi \qquad (2.20)$$

as the reader may straightforwardly check by using a polar coordinate system and a circular trajectory that contains the origin of coordinates. Mathematically, what happens is that the angle is not well defined at the origin (it is multivalued, in fact), and thus the integration does not vanish[8].

---

[7] Although non-superconducting the void does not introduce discontinuity in the electromagnetic fields, and one may apply Stokes theorem.
[8] However, if C does not contain the origin of coordinates one has $\oint_C \nabla\theta \cdot d\ell = 0$.

In conclusion, as long as ω and χ are interpreted as functions related to the super-conductor that admit mathematical 'singularities' at some points, one may expect measurable effects, for instance in the flux threading across the surface that contains such singular points. In more technical terms, according to equation (2.17) the flux of **B** across a given area contains a contribution that is sensitive to the topology of the region.

2.4.2 On the nature of ω and χ: wave functions

As said, the above ideas were already discussed by London in the late 1940s [7]. He introduced the concept of *fluxoid* that is nothing but the function χ (recall that dimensionally χ is a magnetic flux). Giving a step further, he also advanced a property that was not resolved experimentally until the 1960s: the function χ and the related current density ω straightforwardly connect to a quantum mechanical wave function that describes the superconducting charge carriers as a whole. More specifically, he already proposed the relation $\chi = (\hbar/e)\theta$ between the gauge adjusting function χ and the phase of the superconducting wave function $\Psi = \Psi_0 \exp(i\theta)$. With this identification, the multiple valuedness of the exponential in the wave function $(\exp(i(\theta + 2n\pi)) = \exp(i\theta))$ gives way to the well known result

$$\Lambda_{\mathrm{L}} \oint_C \mathbf{J} \cdot d\ell + \iint_S \mathbf{B} \cdot d\mathbf{s} = \frac{\hbar}{e} \oint_C \nabla\theta \cdot d\ell = 2n\pi\frac{\hbar}{e} \equiv n\,\Phi_{\mathrm{U}} \qquad (2.21)$$

Here, following London's terminology, the magnetic flux measured across the surface of a superconducting sample with a hole is quantised in terms of the so-called *universal unit of the fluxoid* $\Phi_{\mathrm{U}} = 2\pi\hbar/e$.

Remarkably, London's prediction was confirmed one decade later, based on measurements of trapped magnetic flux in hollow superconducting samples. Two teams of researchers [10] and [11] verified relation (2.21) with two different materials: tin and lead. As said by London, the measured flux was quantised in terms of one fundamental unit. In passing, both groups obtained a fundamental result. Indeed, the elementary unit, established by London in terms of the elementary charge[9], had to be modified to the expression for the *superconducting flux quantum*:

$$\Phi_{\mathrm{U}} \to \Phi_0 = \frac{\pi\hbar}{e} \qquad (2.22)$$

Outstandingly, this experimental result tells us about the nature of the super-conducting carriers: the elementary constituent holds a charge of $-2e$.

Let us elaborate London's proposal a bit further. To start with, the rationale behind the model (2.14) for the superconducting material law connects naturally with the well established bases of quantum mechanics. In the formulation, electromagnetic potentials acquire a physical significance and give way to measurable effects. Topological features (as a void in our case) provide a means for such measurements. This is exactly implied in the concept of figure 2.2. Owing to the non-local nature of wave functions Ψ,

---

[9] We are using the convention $e = 1.602\ 176\ 62 \times 10^{-19}$ coulombs for the elementary charge. Recall that, dealing with electrons or 'superelectrons', one has to introduce a minus sign!

the interaction of charged particles with an electromagnetic field may be affected by the presence of the void and recorded by measuring the *fluxoid* around such void. Thus, intuiting the quantum nature of the phenomenon, London proposed the relation mentioned above between χ, ω and a global wave function for the superconducting carriers ensemble: $\Psi = \Psi_0 \exp(i\theta)$.[10] Let us go quantitative. To start with, we recall that the fundamental relation concerning the evolution of a quantum system, i.e. Schrödinger's equation for the superconductor

$$i\hbar \frac{\partial \Psi}{\partial t} = \mathcal{H}\Psi \tag{2.23}$$

with $\mathcal{H}$ being the Hamiltonian of the system[11]. First, we recall that starting with this equation, for the free particle Hamiltonian $\mathcal{H} = -\mathbf{p}^2/2m$ with the correspondence rule for the momentum $\mathbf{p} \rightarrow -i\hbar\nabla$ one obtains the probability current density associated to the ensemble described by Ψ:

$$\mathbf{j} = \frac{\hbar}{2mi}\left(\Psi^*\nabla\Psi - \Psi\nabla\Psi^*\right) \tag{2.24}$$

with Ψ^* denoting the complex conjugate of Ψ. For an ensemble of charged particles, $\mathbf{j}$ may be converted into a charge current density just multiplying by some charge q. Let us denote such current density by ω and evaluate it within the above framework. We get

$$\omega \equiv \frac{q\hbar}{2mi}\left(\Psi^*\nabla\Psi - \Psi\nabla\Psi^*\right) = \frac{\hbar q\, \Psi_0^2}{m}\nabla\theta \tag{2.25}$$

By comparing this expression to equations (2.7) and (2.15) one obtains the equivalence relations

$$\begin{cases} \theta \rightarrow \dfrac{q}{\hbar}\chi \\ |\Psi|^2 = \Psi\Psi^* \rightarrow n_s \end{cases}$$

i.e. as said, the fluxoid straightforwardly relates to the phase of the wave function, and the density of particles (superconducting charge carriers) n_s relates to the squared modulus of the wave function.

For completeness, we mention two further aspects concerning the quantum mechanical description of superconductivity. First, we recall that, dealing with charged particles in electromagnetic fields one should in fact modify the Hamiltonian of the system. By using the 'minimal coupling principle'[12] one should indeed use

---

[10] Here, we are assuming a representation in which Ψ_0 is real.

[11] A big amount of excellent textbooks introducing the formalism of quantum mechanics are available. A dedicated introduction/reminder focused on the application to superconducting materials may be found in the book by Orlando and Delin [12].

[12] Both in classical and in quantum mechanics, the principle of minimal coupling accounts for electromagnetic interactions by a simple replacement rule $\mathbf{p} \rightarrow \mathbf{p} - q\mathbf{A}$.

$$\mathcal{H} \xrightarrow{\text{minimal coupling}} -\frac{(-i\hbar\nabla - q\mathbf{A})^2}{2m} \tag{2.26}$$

As the reader may check, if one reproduces the procedure that leads to equation (2.24) for this case, one gets

$$\omega \xrightarrow{\text{minimal coupling}} \mathbf{J} = \omega - \frac{q^2}{m}|\Psi|^2\mathbf{A} \tag{2.27}$$

Recalling the definition of Λ_L and the relation $\Psi\Psi^* \to n$, this is nothing but the restatement of equation (2.14), a fact that leads to an interesting physical interpretation.

> ω *gives the evolution of the superconducting carriers' density in the absence of magnetic fields, while $\mathbf{J}$ represents the full current density under the action of the magnetic field.*

For further application, we end this section by analysing another aspect concerning the Hamiltonian. Strictly speaking the minimal coupling principle has to be applied to the full electromagnetic field, and this concerns the possible inclusion of an electrostatic term. This is done through the rule for the electrostatic part $i\hbar\partial_t \to i\hbar\partial_t - q\phi$, that leads to

$$i\hbar\frac{\partial\Psi}{\partial t} = \frac{(-i\hbar\nabla - q\mathbf{A})^2}{2m}\Psi + q\,\phi\,\Psi \tag{2.28}$$

2.4.3 Ginzburg–Landau equations

Since the very first statement of any book dealing with superconductivity one may guess that thermodynamics will play an essential role. The 'condensation' to the superconducting state by lowering temperature and, reversely, the destruction of superconductivity by heating the sample up just indicate that a complete description of the phenomenon needs to include the concepts of this discipline. The Ginzburg–Landau theory [13] succeeded to bring together the above introduced quantum mechanical description and the needed thermodynamical background.

Landau and Ginzburg issued a theory that describes the superconducting state in thermodynamic equilibrium. The essential ingredient is the so-called complex order parameter, which is basically the above introduced Ψ. Second, the theory takes advantage of a concept that is essential both in thermodynamics and in quantum mechanics, viz., the variational statement of the problem. Thus, they defined an expression of the free energy for the superconductor, $\mathcal{F}_s$, whose minimisation under appropriate constraints allows to determine the state of the system. On the one side, $\mathcal{F}_s$ contains the electromagnetic energy terms that generalise equation (2.9). Physically, they stand for (i) the magnetostatic energy and (ii) the kinetics of the gauge invariant supercurrent (equation (2.27)) On the other side, as the density of superconducting carriers is represented by $\Psi^*\Psi = |\Psi|^2$, $\mathcal{F}_s$ contains terms that account for $|\Psi|^2$ and its gradient. They are related to the self-interaction of the field that describes the superconducting particles. These terms may be understood as the leading part of a power series expansion. Henceforth, in principle, the theory is valid close to the superconducting transition temperature where $|\Psi|^2$ and its variations are expected to be small. Nevertheless, it is well known that many conclusions are successfully extrapolated to low temperatures.

Customarily, the Ginzburg–Landau free energy density[13] is written as follows

$$\mathcal{F}_s = \mathcal{F}_{N0}(T) + \alpha|\Psi|^2 + \frac{\beta}{2}|\Psi|^4 + \frac{1}{2m}\left|\left(\frac{\hbar}{i}\nabla - q\mathbf{A}\right)\Psi\right|^2 + \frac{B^2}{2\mu_0} \qquad (2.29)$$

Here, $\mathcal{F}_{N0}$ is the energy of the normal state in the absence of magnetic fields and α, β are parameters of the theory to be analysed below.

$\mathcal{F}_{N0}$ is introduced so as to account for the *condensation energy* that makes the superconducting state more favourable to the normal state below T_c. Thus, just by comparing the energy of the perfect Meissner state (absolutely no field within the superconductor) with the fully penetrated normal state, one has

$$\mathcal{F}_{N0}(T) \equiv \mathcal{F}_{S0}(T) + \frac{\mu_0 H_c^2}{2} \Rightarrow \mathcal{F}_{S0}(T) = \mathcal{F}_{N0}(T) - \frac{\mu_0 H_c^2}{2} \qquad (2.30)$$

with H_c^2 being the so-called *thermodynamic critical field* that accounts for such a reduction of energy. Notice that the reduction goes to zero for an applied field $H_a = H_c$ as required!

Concerning α and β, following Tinkham [14] one may show that, just by imposing that $\mathcal{F}_s - \mathcal{F}_{N0}$ has a minimum, well within the sample (electromagnetic fields go to zero and Ψ takes the value so-called Ψ_∞) one has

$$\frac{\alpha}{\beta} = -|\Psi_\infty|^2; \quad \alpha = -\frac{\mu_0 H_c^2}{|\Psi_\infty|^2} \qquad (2.31)$$

Notice that $|\Psi_\infty|^2$ represents the uniform density of superelectrons in equilibrium when no field is present and is the higher bound for $|\Psi|^2$. Notice also the negative value of the coefficient α, that acts to reduce energy of the system by 'condensation' to the superconducting state

Apparently, the third term in (2.29) relates to some kinetic energy (it may be written in the form $(\mathbf{p}\Psi)^*(\mathbf{p}\Psi)/2m$), and the fourth term accounts for the magnetostatic energy.

In order to facilitate the physical interpretation, one may expand the third term and compare to the gauge invariant current density $\mathbf{J}$ given by equations (2.25) and (2.27). The reader can check that after some algebra one obtains

---

[13] The full energy is calculated by integrating over the sample's volume.

$$\mathcal{F}_s = \mathcal{F}_{N0}(T) + \alpha|\Psi|^2 + \frac{\beta}{2}|\Psi|^4 + \frac{\hbar^2}{2m}\left(\nabla\sqrt{|\Psi|}\right)^2 + \frac{\mu_0\lambda_L^2}{2}J^2 + \frac{B^2}{2\mu_0} \tag{2.32}$$

i.e. the term corresponding to the quantum mechanical canonical momentum (third term in equation (2.32)) gathers the kinetic contribution related to the current density (also appearing in London's theory (equation (2.9)), but additionally includes a contribution related to inhomogeneity in the distribution of charge carriers ($\nabla\sqrt{|\Psi|}$). This makes a fundamental difference between both approaches. Although London merits the introduction of a superconducting wave function, his theory was developed disregarding the possibility of spatially changing $|\Psi|^2$. By means of equation (2.32) one has the possibility of investigating a new set of physical properties. As it will be seen later, this opened a new era in superconductivity. By now, let us take a closer look at equation (2.32). Regarding the free energy as a functional that depends on the fields Ψ and $\mathbf{A}$ one[14] obtains the celebrated set of Ginzburg–Landau differential equations by minimising with respect to such variables. By applying the standard minimisation techniques[15]

$$\text{Min } \mathcal{F}_s[\Psi(\mathbf{r}), \mathbf{A}(\mathbf{r})] \Rightarrow \begin{cases} \dfrac{\partial\mathcal{F}_s}{\partial\Psi^*} = \sum_j \dfrac{\partial}{\partial x_j}\dfrac{\partial\mathcal{F}_s}{\partial(\partial\Psi^*/\partial x_j)} \\ \dfrac{\partial\mathcal{F}_s}{\partial\mathbf{A}} = \sum_j \dfrac{\partial}{\partial x_j}\dfrac{\partial\mathcal{F}_s}{\partial(\partial\mathbf{A}/\partial x_j)} \end{cases}$$

and after some algebra, these equations become, respectively

$$\begin{cases} \alpha\Psi + \beta|\Psi|^2\Psi + \dfrac{1}{2m}\left(\dfrac{\hbar}{i}\nabla - q\mathbf{A}\right)^2\Psi = 0 & \text{(First GL equation)} \\ \mathbf{J} = \dfrac{q\hbar}{2mi}(\Psi^*\nabla\Psi - \Psi\nabla\Psi^*) - \dfrac{q^2}{m}|\Psi|^2\mathbf{A} & \text{(Second GL equation)} \end{cases} \tag{2.33}$$

i.e. one obtains the law for the spatial variation of the order parameter and consistently recovers the definition of the electrical current density in a superconductor.

2.5 The characteristic lengths

It is commonly said that the Ginzburg–Landau (GL) equations contain the two characteristic lengths of superconductivity, essential parameters of the theory, that determine the behaviour of the material. Equation (2.32) transparently displays the first of them, i.e. λ_L whose physical interpretation is already clear, as it was already analysed in section 2.3 in the realm of London's theory. It is the characteristic scale for the variation of the magnetic field within the superconductor. As the GL theory incorporates the spatial dependence of the carrier density (modulus of the order parameter) one may

---

[14] The role of boundary conditions will not be considered at this step.
[15] The well known Euler–Lagrange equations in field theory. A brief survey on this topic is given in section 5.1.

advance that an analogous parameter will exist, related to the variation of Ψ. In order to see how this arises in the theory, one may transform the first GL equation in (2.33) by replacing α and β, and introducing the 'normalised' order parameter $f \equiv \Psi/|\Psi_\infty|$ for the case $\mathbf{A} = 0$. One gets:

$$f - f|f|^2 + \frac{\hbar^2}{2m|\alpha|}\nabla^2 f = 0 \tag{2.34}$$

Apparently, the prefactor of the Laplacian term has the dimensions of a squared length, that one may define by

$$\xi \equiv \frac{\hbar}{\sqrt{2m|\alpha|}} \tag{2.35}$$

and is named after *coherence length* of the material. It establishes the scale of variation of the order parameter.

At the phenomenological level of the GL theory one may not predict values either for λ_L or ξ. For this purpose, one should introduce the microscopic theory. As this is out of our scope, we shall adopt the *phenomenological* point of view: describe the properties of our material of interest relying on the knowledge of these parameters based on experimental information.

Perhaps, the most important fact related to the characteristic lengths is the classification of superconducting materials in terms of the celebrated Ginzburg–Landau parameter

$$\kappa \equiv \frac{\lambda_L}{\xi} \tag{2.36}$$

As it is argued in most textbooks on superconductivity, by recalling the concept of *surface energy*, one may give a qualitative explanation for the classification of superconductors in either type-I or type-II according to the values of κ. Thus, according to figure 2.3, that pictures the penetration of magnetic field in a semi-infinite medium occupying the region $x > 0$:

- For $\lambda_L < \xi$, there is a region from which the magnetic field must be excluded, but which is not accompanied by a reduction of energy provided by the condensation to the optimum value of the order parameter $|\Psi_\infty|$. This fact disfavours the appearance of boundaries containing this energetical barrier[16].
- For $\lambda_L > \xi$, the formation of boundaries is favoured, because there the reduction of energy due to the prompt appearance of condensation energy is advantageous.

Certainly, the above argumentation is schematic, and does not allow a quantitative determination of the crossover between one kind of behaviour and the other. The topic is by no means simple. In fact, as it is well known, although it was Landau

---

[16] Recall the factor $\alpha|\Psi|^2$ in the free energy.

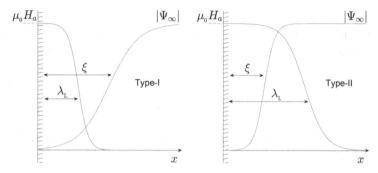

Figure 2.3. Schematic representation of the decay of the magnetic field and order parameter at the interface between the superconductor and a normal zone in a semi-infinite medium. The magnetic field decays from the surface towards zero with the characteristic length λ_L, while the absolute value of order parameter increases from zero to the maximum value $|\Psi_\infty|$ well within the superconductor, with the characteristic length ξ. The situations for a type-I and a type-II superconductor are distinguished.

himself who already suggested that a separation between both kinds of behaviours would be determined by the precise condition

$$\kappa > \frac{1}{\sqrt{2}}, \tag{2.37}$$

the ultimate elaboration of the theory for the behaviour of superconductors with different values of κ in a magnetic field is due to Abrikosov [15], who made the quantitative study of the consequences of having $\kappa > 1/\sqrt{2}$ in the GL free energy. It is since his work that the classification of superconductors in type-I materials ($\kappa < 1/\sqrt{2}$) and type-II materials ($\kappa > 1/\sqrt{2}$) is clear.

Among other results, Abrikosov's calculations could explain the unexpectedly high values of the observed critical fields in some experiments with superconducting alloys. This had remained a puzzle since the 1930s. The main physical novelty in his theory is that in some superconductors partial penetration of **B** in the form of quantised magnetic flux tubes (Abrikosov vortices) may occur. Such vortices have the following properties:

- The related superconducting currents flow in circular paths surrounding the so-called core of the vortex, a region of size 2ξ, where $|\Psi|$ goes to 0 from the value $|\Psi_\infty|$ at large distances.
- According to the quantisation condition (equation (2.21)), that applies around voids in the superconductor, they enclose an amount of flux Φ_0.
- By allowing the penetration of magnetic flux in an ordered array of such vortices, the reduction of energy due to the condensation term in the boundaries surrounding such tubes allows to have an equilibrium state in very high applied magnetic fields. Though partially diamagnetic, the material is still superconducting.

In his work, Abrikosov evaluated with remarkable success the properties that had been experimentally determined for several superconducting alloys. In particular, he calculated the characteristic fields that define the range for which the superconductor develops the above described structure of penetrating flux tubes (see figure 2.5). These are the so-called *lower critical field* H_{c1} and *upper critical field* H_{c2}. The interested reader may find instructive derivations of these quantities in [12] and [14] by using the GL equations and convenient approximations. They read

$$H_{c1} = \frac{\Phi_0}{4\pi\mu_0\lambda_L^2}K_0(1/\kappa) \tag{2.38}$$

$$H_{c2} = \frac{\Phi_0}{2\pi\mu_0\xi^2} \tag{2.39}$$

According to the above concepts one may now interpret H_{c1} as the field for which it is energetically advantageous that the first vortex nucleates within the sample, i.e. the increase of magnetostatic energy is compensated by the reduction through condensation energy, and H_{c2} identified with the highest value of applied field for which a solution of the first GL equation may exist in the conditions enabled by a very small value of the order parameter (that goes to 0 at $T = T_c$). As shown in figure 2.7, both H_{c1} and H_{c2} are dependent on temperature through the characteristic lengths $\lambda_L(T)$ and $\xi(T)$.

2.6 Flux vortices

The presence of flux vortices determines the rich phenomenology that characterises the behaviour of type-II superconductors. In this section we review some properties of their actual structure, mutual interactions and relation with the background (material structure, external electromagnetic sources, and so on). A basic knowledge is crucial for the understanding of these materials in the customary application range.

2.6.1 Structure of an isolated vortex

To start with, we analyse the structure of an isolated vortex within an ideally infinite superconductor. This avoids technical complications due to the boundary conditions. The application of equation (2.33) allows to obtain the dependence of the fields $\Psi(\mathbf{r})$ and $\mathbf{B}(\mathbf{r})$ by using some assumptions. Thus, considering the cylindrical symmetry with the z-axis identified with the direction of the magnetic field, we use

$$\Psi(\mathbf{r}) = f(r)e^{i\theta}; \quad \mathbf{b}(\mathbf{r}) = b_z(r)\hat{\mathbf{k}} \tag{2.40}$$

Here, we recall that the lowercase for the magnetic field has been introduced so as to enable further simplicity in the notation[17].

---

[17] Eventually, a sort of 'molecular' theory with averages over thousands of vortices will be used. Our fields of interest will be mean values that smooth fluctuations of individual vortex fields: $\mathbf{B}(\mathbf{r}) \equiv \langle\mathbf{b}(\mathbf{r}_v)\rangle$. In general, we prefer to use $\mathbf{B}(\mathbf{r})$ for the magnetic field of our problem, whatever the length scale of interest, unless different scales are involved as in the previous formula.

Even within the above simplification, the GL equations must be solved numerically. A useful quasi-analytical approximation, valid for the whole range $\kappa > 1/\sqrt{2}$ was found by Clem [16][18]. On the other hand, as argued in many superconductivity textbooks, fully analytical approximations may be found for the high-κ region, based on the London limit of the GL equations. Related to this assumption, figure 2.4 may help to visualise the basic facts:

- For $\kappa \gg 1$ one may replace the order parameter by a step function that jumps from 0 at $r < \xi$ to the maximum value $|\Psi_\infty|$ for $r > \xi$. Then, $b_z(r)$ is obtained by solving London's equation (2.5) for $r > \xi$ with the boundary condition that it remains constant for $r < \xi$. One gets

$$b_z(r) = \begin{cases} b_0, & r < \xi \\ b_0 K_0(r/\lambda_{\mathrm{L}}), & r > \xi \end{cases} \tag{2.41}$$

- It may be shown that the value of the magnetic field at the centre of the vortex satisfies $b_0 \approx 2\mu_0 H_{c1}$.

For our purposes, it suffices to say that (mostly) practical superconductors fall in the category of high-κ materials. Thus, although the exact picture is a bit more complicated due to the influence of other factors as the anisotropy both of λ_{L} and ξ, one may build the model for the mixed state of such materials based on the picture to the right of figure 2.4, i.e. vortices may be considered as tiny tubes of magnetic flux (of typical radius λ_{L}) within a uniform fully superconducting medium. This will be the building block for the forthcoming steps of the theory.

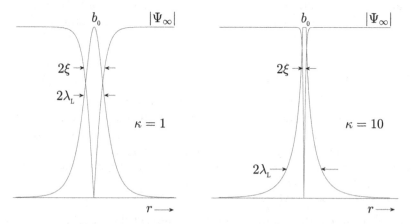

Figure 2.4. Variation of the magnetic field and order parameter for an isolated vortex well within a type-II superconductor. Left: corresponding to the value of the GL parameter $\kappa = 1$. Right: same for $\kappa = 10$.

---

[18] The main result in that paper, i.e. $b_z \simeq K_0(\sqrt{r^2 + \xi^2}/\lambda_{\mathrm{L}})$ for $\kappa \approx 1$ has been used in order to generate figure 2.4.

2.6.2 Interaction between vortices

When the applied magnetic field exceeds H_{c1}, more vortices are nucleated in the sample. Eventually, an equilibrium configuration is established, characterised by an ensemble of flux tubes at precise locations. The actual disposition may be qualitatively described by the essential fact that two of such vortices interact through a repulsive force along the direction between themselves. As shown in figure 2.5, when 'pressed' inwards by the applied magnetic field they settle up in a perfect triangular lattice[19]. For high-κ materials at low densities of vortices $(H_a \gtrsim H_{c1})$ such a picture may be obtained by simple superposition of the structure described by equation (2.41) (and relying on the established knowledge that the full GL energy of the system is minimised by the triangular arrangement! [14, 15]).

The above mentioned repulsion force between vortices may be quantified by applying the famous relation

$$\mathbf{f}_2 = \mathbf{j}_1(\mathbf{r}_2) \times \Phi_0 \hat{k} \tag{2.42}$$

that gives the force per unit length on vortex '2' at position $\mathbf{r}_2$ due to the current density created at such point by vortex '1'.

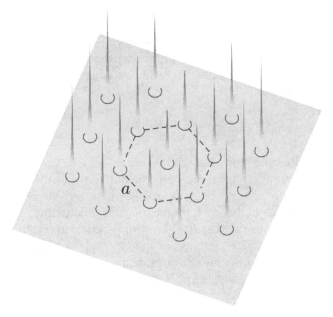

Figure 2.5. Equilibrium triangular vortex lattice. The picture shows the magnetic field (along the z-axis) $b_z(x, y)$ as calculated through equation (2.41) for each individual vortex in the low density limit $a \gg \lambda_L$.

---

[19] In the absence of inhomogeneities, anisotropy or finite size effects of any kind.

We want to stress that the concept of force, as well as the actual expression (2.42) are derived from the 'mechanical' relation connecting forces and variation of energy. Thus, in order to obtain (2.42) one proceeds by taking a derivative of the interaction energy term in the London limit of the free energy, i.e. writing equation (2.9) for $\mathbf{b} = \mathbf{b}_1 + \mathbf{b}_2$ and using $\mathbf{f}_2 = -\nabla_2 U_{\text{INTER}}$ on the part of U that excludes self interactions.

As remarked by Chen and co-workers [17], an important conceptual flaw in the interpretation of equation (2.42) must be taken into account because it has lead to frequent errors. Although the expression is formally analogue to the magnetostatic term of the 'Lorentz force', in fact it is mostly related to the kinetic part in the free energy, which is higher than the magnetostatic contribution. Thus, a more appropriate term would be *Lorentz-like* force. As a matter of fact, the former is just the opposite, because it would give the force on the current due to the magnetic field and not *vice versa*.

2.6.3 The vortex lattice: average of *b*

By further exploiting the 'mechanical' interpretation of equation (2.42), i.e. flux vortices may be considered as 'particles' subjected to interaction forces, one may describe the equilibrium vortex lattice as such a configuration for which the total force on each vortex goes to zero by compensation of the different contributions. As said, this is achieved through the highly symmetric situation shown in figure 2.5. Recall that each vortex is surrounded by six nearest neighbours at equal distances of value a.

Owing to the characteristic values of the involved distances in experiments with bulk samples, we will hereafter work with averaged values in the following sense. According to figure 2.5 for the triangular lattice arrangement shown there, one may introduce the flux per unit area

$$\frac{\Phi}{S} = \frac{\Phi_0}{\sqrt{3}\, a^2/2} \tag{2.43}$$

Then, for such situations for which the experimental resolution is much above a, one may define a continuous 'local' magnetic field in terms of this average quantity. This field will be more intense in those zones where the intervortex distance is smaller[20]. Thus, although the real profile of b is fluctuating, in a scale with macroscopic resolution one may use $B = \Phi S$. For practical application, as indicated in figure 2.6, we will consider the 'macroscopic cell' approximation in which one may discriminate between different locations, being representative of points around which different densities of vortices may occur. Notationally

$$B(x, y) = \frac{2\Phi_0}{\sqrt{3}\, a^2(x, y)} \equiv n_{\text{v}}\Phi_0 \tag{2.44}$$

with n_{v} being the mean number of vortices per unit area at the 'mean point' (x, y).

---

[20] Recalling that $\Phi_0 = 2.07 \times 10^{-15}$ Wb, for typical field values in the range of 1T, a is a value around $\simeq 100$ nm.

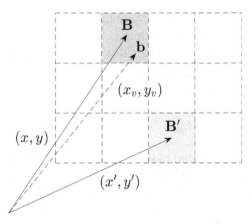

Figure 2.6. Macroscopic cell approximation. Different average flux densities may occur around the points (x, y) and (x', y'). The position of an 'individual vortex' is indicated by (x_v, y_v) and its field by **b**.

Although in this chapter, the distinction between b and B is fundamental, in the rest of the book, we will drop the notation for simplicity. Being concentrated on macroscopic phenomena, it is implicit that our variables are (x, y) and B, though not explicitly indicated.

2.6.4 Equilibrium magnetisation curve of a type-II superconductor

As shown above, by taking advantage of Abrikosov's result one may deduce a number of properties about the behaviour of type-II materials for applied fields H_a slightly above H_{c1}, by using the London approximation for the free energy and assuming the existence of a lattice of vortices. However, the overall response to the applied magnetic field can only be studied by the minimisation of (2.32), which in general is by no means a simple task. In fact, it must be done by resorting to numerical methods. On the other hand, such information is essential for the analysis of a fundamental experimental quantity, i.e. the so-called volume magnetisation defined in (1.23). Indeed, the full curve $(M_v(H_a))$ contains much physical information.

A pedagogical approach to the interpretation of the general behaviour may be found in [14], relying on the analysis of different regimes that may be essentially joined so as to cover the full range. Figure 2.7 has been compiled based on this. Notice that we use the notation $\mathbf{M}_{EQ}$ as an indication that, for each value of the magnetic field, the sample settles at the (unique) equilibrium configuration predicted by (2.32). We recall that $\mathbf{M}_{EQ}$ is evaluated after solving for the fields **A** and Ψ and using

$$\mathbf{M}_{EQ} = \frac{\langle \mathbf{B} \rangle}{\mu_0} - \mathbf{H}_a \qquad (2.45)$$

with $\langle B \rangle$ indicating the averaged magnetic field over the full superconducting volume.

In the figure, one may notice the appearance of several regimes:

(A) The linear Meissner state region $M_{EQ} \approx -H_a$ for $H_a < H_{c1}$.

(B) A low field region with a sudden drop of M_{EQ} with the nucleation of the first vortices ($\langle B \rangle \propto \{\ln[3\Phi_0/4\pi\mu_0(H_a - H_{c1})]\}^{-2}$).

(C) A region of intermediate flux densities that may still be predicted by neglecting the cores of the vortices ($M_{EQ} \propto \ln(H_{c2}/H_a)$).

(D) A linear behaviour $M_{EQ} \propto H_a - H_{c2}$ close to the upper critical field.

Some quantitative aspects that may be deduced from figure 2.7 together with the expressions for H_{c1} and H_{c2} will be of interest for our purposes. Notice that the maximum (absolute) value of the magnetisation is $M_{EQ}^* = M_{EQ}(H_a = H_{c1})$, which decreases with the value of κ. Thus, high-κ materials (as is the case of high-T_c oxides, with typical values of $\mu_0 H_{c1} \approx 1$ mT) will have rather small values of equilibrium magnetisation. On the other hand, they will remain superconducting up to high values of magnetic field, because H_{c2} will be large (up to $\mu_0 H_{c2} \approx 100$T for the oxides). Recall expressions (2.38) and (2.39).

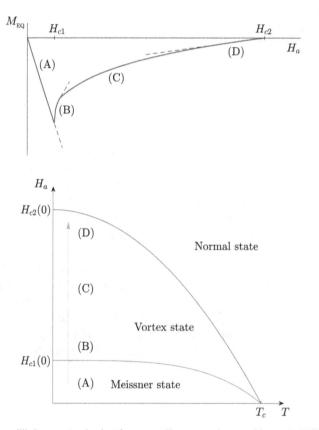

Figure 2.7. Upper: equilibrium magnetisation for a type-II superconductor with $\kappa = 2$. Different regions are defined. A: Meissner state, B: low flux density (distant vortices), C: medium flux density, D: linear behaviour close to H_{c2}. The plot is not to scale. Lower: superconducting phase diagram with typical temperature dependencies $H_{c1}(T)$, $H_{c2}(T)$ plotted with help of equations (2.38) and (2.39).

2.6.5 Vortex dynamics: dissipation

At first glance, according to the above paragraph, high-κ materials appear as natural candidates for large scale applications due to the high values of H_{c2}. However, the real facts are not that simple. On the one hand, many applications rely on non-equilibrium states due to the application of transport currents additional or alternative to the application of H_a (superconducting cables), a fact that has not been included in the theory. On the other hand, another important sector of applications demand a high value of magnetisation (levitation systems) that is precluded by the smallness of H_{c1}. Below, we analyse this situation. In fact, the solution to the first problem will also fix the second.

As said above, the Abrikosov vortex lattice consists of an equilibrium situation for which flux vortices settle at special positions that give way to a vanishing net force on each of them. However, one may also argue that if a transport current $\mathbf{J}_\mathrm{T}$ is superimposed on this configuration with current loops around each vortex, the i-th vortex will suffer a net force given by

$$\mathbf{f}_i = \mathbf{J}_\mathrm{T} \times \Phi_0 \hat{k} \tag{2.46}$$

Then, equilibrium will break down and the vortex will move along the lines of $\mathbf{f}_i$. As it is well known, this will cause dissipation. The underlying physics is as follows. The vortex core is normal conducting[21] ($\Psi(r \to 0) \to 0$) and if it goes through the region with current flow, one has an effective dissipative medium. This picture is on the basis of the celebrated Bardeen–Stephen's model [18]. These authors showed that the electric fields generated by the movement of the vortex drive the current across the core and, thus, $\mathbf{J}_\mathrm{T}$ will contain a normal conducting contribution. Without going into the details of the derivation, we just show the main result of this model

$$\rho_\mathrm{f} = \frac{B}{\mu_0 H_{c2}} \rho_\mathrm{n} \tag{2.47}$$

i.e. the so-called *flux flow* resistivity of the superconductor subject to a transport current is proportional to the *normal state* resistivity and to the magnetic field.

In conclusion, the application of a transport current, however small, will lead to dissipation, essentially described by equation (2.47). By straightforward application of Faraday's law to the moving vortices, the reader may show that the induced electric field is parallel to the current density, thus giving place to a loss of energy.

2.6.6 Vortex dynamics: flux pinning

Apparently, if a type-II superconductor is to be useful under transport current, one should provide a mechanism that blocks the continuous drift of vortices. This longly studied concept is known as *flux pinning* of vortices, and a number of solutions have been well known for decades. Material scientists have succeeded to pin vortices by natural or artificially created imperfections such as voids, grain boundaries, irradiation

---

[21] In fact, the non-superconducting behaviour extends beyond the very centre, to a region of basic size $r \lesssim \xi$.

defects, inclusions, etc, typically of the size of the vortex so as to be effective. For our analysis, we will just consider that, in the presence of pinning, vortices may stay in an induced equilibrium state characterised by the condition $\mathbf{f}_i + \mathbf{f}_p = 0$ with $\mathbf{f}_i$ the Lorentz-like force and $\mathbf{f}_p$ characterising the interaction with the defective structure.

We notice that the existence of $\mathbf{J}_T$ will have a relevant consequence concerning the magnetic field picture within the superconductor. Let us analyse a model system consisting of a superconducting slab along which a macroscopically uniform current density is flowing[22]. Here, the term 'macroscopically uniform' is used in the sense of averages discussed around equation (2.44). From the macroscopic point of view, by applying Ampère's law, one may deduce that a uniform current implies a uniform gradient of magnetic field. The actual underlying situation may be understood as pictured in figure 2.8. The infinite superconducting slab is defined by $|y| \leqslant w$ with $\mathbf{J}_T = J_0 \hat{\imath}$. Within the slab one has[23]

$$\mu_0 J_0 = \frac{dB_z}{dy} \quad \Rightarrow \quad B_z = \mu_0 J_0 \, y, \quad |y| < w \tag{2.48}$$

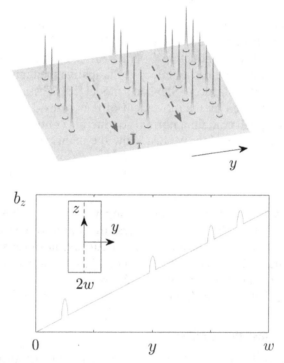

Figure 2.8. (Schematic) Superconducting slab with uniform transport current along the x-axis. Upper: rows of induced vortex lines for $0 < y < w$. Lower: representation of the 'mesoscopic' magnetic field in that region.

---

[22] In section 3.3 we will show that this is a rather realistic situation for type-II superconductors in practical applications.
[23] Outside, the field created by the current $\mathbf{B}$ remains constant: $B_z = \text{sign}(y)\mu_0 J_0 w$.

However, as shown schematically in the plot, the real situation consists of a mesoscopically fluctuating field with rows of vortices more and more densely packed as B_z increases. These vortices have been nucleated by action of the applied current, but they remain static due to the pinning force. Outstandingly, this picture tells us that, enabled by the possibility of packing vortices with a high slope close to the surface of the sample, and leaving a flux free core, one may produce high magnetic moments, and thus a volume magnetisation much beyond M_{EQ}, as advanced above[24].

2.7 Representative superconducting materials

Several classification criteria may be used with superconducting materials, in order to accommodate the already large number of compounds for which superconductivity has been reported. Thus, as said before, (i) the behaviour in a magnetic field determines the basic scheme introduced by Abrikosov of type-I and type-II materials. Also, (ii) one may speak about low temperature superconductors (LTS or low-T_c) and high temperature superconductors (HTS or high-T_c), usually according to the boundary established in 30 K. (iii) According to the material structure, one has superconductivity in pure metals, alloys, superconducting ceramics that include the high T_c copper oxides and magnesium diboride among others. In passing, we note that superconductivity under extreme experimental conditions has also been reported (though yet far from understood)[25].

Below, just for the reader's convenience when performing basic evaluations for some of the most popular materials, we offer a list of parameter values in a selection of representative compounds. Recall that, in general, pure metals are type-I, that alloys and ceramics are type-II (with very few exceptions as the case of Nb). Notice also that the high $-T_c$ superconductors are anisotropic materials, and their parameters are given either in the a, b plane (CuO planes) or along the c axis (perpendicular to CuO planes). This feature should be kept in mind even when bulk samples are considered, because material processing techniques are typically used and the microscopic anisotropy may be propagated to the macroscopic level.

The materials listed below are very different in nature and so do the microscopic interactions that produce superconductivity. However, the phenomenological framework introduced in this chapter has been proved to be valid for all cases.

Low $-T_c$ superconductors (type-I).

Material (discovery)	T_c (K)	$\mu_0 H_c$ (T)	λ_L (nm)	ξ (nm)
Sn (1913)	3.72	0.031	50	230
Pb (1913)	7.20	0.08	40	90

Source: values taken from [12].

---

[24] The physical counterpart will be described in section 3.3.

[25] As the recent discovery of superconductivity in a carbonaceous sulphur hydride at 280 K and 271 GPA [19].

Low $-T_c$ superconductors (type-II).

Material (discovery)	T_c (K)	$\mu_0 H_{c2}$ (T)	λ_L (nm)	ξ (nm)
NbTi (1962)	9.5	13	300	4
Nb$_3$Ge (1973)	23	38	90	3

Source: values taken from [12].

High-T_c superconductors (type-II).

Material (discovery)	T_c (K)	$\mu_0 H_{c1}$ (T)	$\mu_0 H_{c2}$ (T)	λ_L (nm)	ξ (nm)
YBa$_2$Cu$_3$O$_7$ (1987)	95	≈0.1	>100	30; 200	3; 4
MgB$_2$ (2001)	39	≈0.03	≈10	≈130	≈5

Source: values taken from [20–22].

2.8 Review exercises

Review exercise R2-1

Make the necessary operations in order to obtain equation (2.11) from the stationarity condition of the energy as given in equation (2.9). This entails to compute explicitly the partial derivatives respect to the field and its derivatives, i.e. those defined in equation (2.10) and considering the appropriate boundary conditions.

Hint: it may be useful to consider that the superconductor is embedded in a medium described by an infinite penetration depth, i.e. $\lambda \rightarrow \infty$.

Review exercise R2-2

Obtain the **J**, **A** formulation of equation (2.9). In chapters 9–12 this will be used to evaluate flux penetration profiles in superconductors by application of variational principles. Change variable **B** to **A** and **J**.

Review exercise R2-3

(*Integration around the void*) Make explicit integration of the gradient of the polar angle, both around the circular path, the square path and also along a path (which you may choose) that does not include the singularity.

Hint: use a polar coordinate system.

Review exercise R2-4

Starting with Schrödinger's equation for the wave function Ψ of a free particle

$$i\hbar\frac{\partial\Psi}{\partial t} = \mathcal{H}\Psi = \frac{\mathbf{p}^2}{2m}\Psi \qquad (2.49)$$

and using the minimum coupling principle $\mathbf{p} \rightarrow \mathbf{p} - q\mathbf{A}$ derive equation (2.27), that gives the expression for the probability current density in the presence of an electromagnetic field

---

Hint: it may be useful to consider the complex conjugate of Schrödinger's equation and combine it with the original so as to obtain the time derivative of the probability density $P \equiv \Psi^*\Psi$. Then use the continuity condition $\partial_t P + \nabla \cdot \mathbf{J} = 0$ so as to identify $\mathbf{J}$.

Review exercise R2-5

Reproduce figure 2.4 and analyse the validity of equation (2.41) for representing the profile of the vortex in the high $-\kappa$ limit.

Hint: Compare the results to Clem's model in [16] that approximates the function $\mathbf{b}(\mathbf{r}_v)$ for the full range of values $\kappa > 1/\sqrt{2}$.

References

[1] Kamerlingh Onnes H 1911 Further experiments with liquid helium. C. On the change of electric resistance of pure metals at very low temperatures, etc. IV. The resistance of pure mercury at helium temperatures *Comm. Phys. Lab. Univ. Leiden* **120b**

[2] Kamerlingh Onnes H 1911 Further experiments with liquid helium. D. On the change of electric resistance of pure metals at very low temperatures, etc. V. The disappearance of the resistance of mercury *Comm. Phys. Lab. Univ. Leiden* **122b**

[3] Kamerlingh Onnes H 1911 Further experiments with liquid helium. G. On the electrical resistance of pure metals, etc. VI. On the sudden change in the rate at which the resistance of mercury disappears *Comm. Phys. Lab. Univ. Leiden* **124c**

[4] Meissner W and Ochsenfeld R 1933 Ein neuer effekt bei eintritt der supraleitfähigkeit *Naturwissenschaften* **21** 787–8

[5] Gorter C and Casimir H 1934 Zur thermodynamik des supraleitenden zustandes *Physik Z* **1934** 963

[6] London F and London H 1935 The electromagnetic equations of a supraconductor *Proc. Roy. Soc. Lond.* A **149** 71–88

[7] London F 1950 *Superfluids* vol 1 (New York: Wiley)

[8] London F 1948 On the problem of the molecular theory of superconductivity *Phys. Rev.* **74** 562–73

[9] Bardeen J, Cooper L N and Schrieffer J R 1957 Theory of superconductivity *Phys. Rev.* **108** 1175–204

[10] Deaver B S and Fairbank W M 1961 Experimental evidence for quantized flux in super-conducting cylinders *Phys. Rev. Lett.* **7** 43–6

[11] Doll R and Nabauer M 1961 Experimental proof of magnetic flux quantization in a superconducting ring *Phys. Rev. Lett.* **7** 51–2

[12] Orlando T P and Delin K A 1991 *Foundations of Applied Superconductivity* (Reading, MA: Addison-Wesley)

[13] Ginzburg V L and Landau L D 1950 On the theory of superconductivity *Zh. Eksp. Teor. Fiz.* **20** 1064–82

[14] Tinkham M 1996 *Introduction to Superconductivity International Series in Pure and Applied Physics* (New York: McGraw-Hill)

[15] Abrikosov A A 1957 On the magnetic properties of superconductors of the second group *Sov. Phys. JETP* **5** 1174

[16] Clem J R 1975 Simple model for the vortex core in a type-II superconductor *J. Low Temp. Phys.* **18** 427–34

[17] Chen D-X, Moreno J J, Hernando A, Sánchez A and Li B-Z 1998 Nature of the driving force on an Abrikosov vortex *Phys. Rev.* B **57** 5059–62

[18] Bardeen J and Stephen M J 1965 Theory of the motion of vortices in superconductors *Phys. Rev.* **140** A1197–207

[19] Snider E, Dasenbrock-Gammon N, McBride R, Debessai M, Vindana H, Vencatasamy K, Lawler K V, Salamat A and Dias R P 2020 Room-temperature superconductivity in a carbonaceous sulfur hydride *Nature* **586** 373

[20] Rani P, Hafiz A K and Awana V P S 2018 Temperature dependence of lower critical field of YBCO superconductor *AIP Conf. Proc.* 1953 120026

[21] Sekitani T, Matsuda Y H and Miura N 2007 Measurement of the upper critical field of optimally-doped $YBa_2Cu_3O_7$-δ in megagauss magnetic fields *New J. Phys.* **9** 47

[22] Buzea C and Yamashita T 2001 Review of the superconducting properties of MgB_2 *Supercond. Sci. Technol.* **14** R115–46

IOP Publishing

Macroscopic Superconducting Phenomena
An interactive guide
Antonio Badía-Majós

Chapter 3

Idealised models and equations: examples

This chapter focuses on the resolution of actual situations that illustrate the basic superconducting phenomena introduced in chapter 2. The statements are simplified to some degree so as to allow the analytical solution through relatively simple mathematical constructions. Ready to use expressions are compiled and discussed so as to deal with: (i) the expulsion of magnetic flux (Meissner state), (ii) the transition to zero resistance, (iii) the so-called *critical state model* for type-II superconductors with pinning, and (iv) some basic relations concerning the macroscopic manifestation of the quantum nature of superconductivity (Josephson effect).

By solving the fundamental equations in specific problems (superconducting sphere, thin wire, infinite slab and elementary Josephson junction), we put the focus on allowing to go further than the qualitative understanding provided by the more academic exposition of chapter 2. The selected models are certainly simplifications of real problems. Nevertheless, they give us the chance to further penetrate the phenomenology and provide a tool for quantitative approximations to real life experiments. They are intended to serve as a bridge between the more abstract concepts and the specialised methods introduced in part III. Expectedly, this will help to provide the physical insight which could be somehow shaded behind the technical details of the otherwise necessary numerical techniques.

3.1 Flux expulsion: Meissner state of a sphere

As explained in section 2.2, the second London equation (2.5) accounts for the magnetic flux expulsion phenomenon observed by Meissner and Ochsenfeld [1]. In its most simplified expression, i.e. uniform field applied to a semi-infinite superconducting medium (e.g.: $\mu_0 \mathbf{H}_a = B_0 \hat{\boldsymbol{k}}$ and the superconductor defined by $x > 0$) one has

$$\frac{d^2 B_z}{dx^2} = \frac{B_z}{\lambda_L^2} \quad \Rightarrow \quad B_z(x) = B_0 e^{-x/\lambda_L} \tag{3.1}$$

doi:10.1088/978-0-7503-2711-4ch3

Apparently, this equation allows to interpret λ_L as the parameter that establishes the typical decay distance of the field for a given superconductor at given conditions (in particular one has $\lambda_L(T)$ with a dependence that tells about the superconducting coupling mechanism [2]). However, being oversimplified, the model loses important information as the magnetic moment to be expected in experiments with a real (finite size) sample. Below, we will provide an example on how one can go a bit further for a geometry that may still be solved exactly, i.e. a spherical superconductor under a uniform applied field. Other realistic configurations as those including cylindrical samples and/or non-uniform magnetic fields are better solved by numeric approximations as those in chapters 9, 10, 13 and 14.

Superconducting sphere in a magnetic field

Let us suppose that a superconducting sphere in the Meissner state is exposed to a uniform magnetic field $\mu_0 \mathbf{H}_a = B_0 \hat{\mathbf{k}}$. Let the sphere be of radius a and let its centre be the origin of a spherical coordinate system (r, θ, φ). The solution of this problem may be found in the book by F London [3]. Here, we will obtain it by a slightly different method. Starting with equation (2.14) for $\omega = 0$, one gets the second London equation in terms of the vector potential

$$\nabla \times \nabla \times \mathbf{A} = -\frac{\mathbf{A}}{\lambda_L^2}, \tag{3.2}$$

which is equivalent to equation (2.5) for $\mathbf{B}$.

Solving (3.2) together with $\nabla \times \nabla \times \mathbf{A} = 0$ outside the sphere will provide the full set of physical properties of the problem. To start with, owing to the symmetry of the statement, one may choose the form $\mathbf{A} = (0, 0, A_\varphi(r, \theta))$ for the searched solution in the spherical coordinate system. Recall that this implies the structure $\mathbf{J} = (0, 0, J_\varphi(r, \theta))$ of current density flowing within the sphere.

By using separation of variables, i.e. assuming $A_\varphi(r, \theta) \equiv R(r)\Theta(\theta)$ and disregarding mathematically valid but physically unacceptable diverging solutions one gets the following expression for the vector potential[1]

$$A_\varphi = \begin{cases} A_1\, i_1\!\left(\dfrac{r}{\lambda_L}\right) \sin\theta, & r \leqslant a \\[2ex] a_1 \dfrac{\sin\theta}{r^2} + \dfrac{B_0 r}{2} \sin\theta, & r > a \end{cases} \tag{3.3}$$

where we have used the function[2]

$$i_1(x) = \frac{\cosh(x)}{x} - \frac{\sinh(x)}{x^2}$$

---

[1] After imposing the asymptotic behaviour at large distances ($\mathbf{B}(r \to \infty) \to \mathbf{B}_0$) and continuity of A_φ and B_θ through the surface of the sphere.
[2] A so-called *modified spherical Bessel function* as given by $i_n(x) = \sqrt{\pi/2x}\; I_{n+1/2}(x)$ with $I_{n+1/2}$ a modified Bessel function of the first kind.

and the definition of the coefficients

$$A_1 = \frac{3B_0 a}{2 \sinh(a/\lambda_L)}$$

$$a_1 = \frac{B_0 a^3}{2}\left[3\frac{\lambda_L}{a}\coth\left(\frac{a}{\lambda_L}\right) - 3\left(\frac{\lambda_L}{a}\right)^2 - 1\right]$$

Relying on the above expressions (and by taking $\nabla \times \mathbf{A}$), one may evaluate the overall magnetic field $\mathbf{B} = (B_r, B_\theta, 0)$.

$$B_r = \begin{cases} 2A_1\left[\dfrac{\cosh(r/\lambda_L)}{r/\lambda_L} - \dfrac{\sinh(r/\lambda_L)}{(r/\lambda_L)^2}\right]\dfrac{\cos\theta}{r}, & r \leqslant a \\[4mm] 2a_1\dfrac{\cos\theta}{r^3} + B_0\cos\theta, & r > a \end{cases} \tag{3.4}$$

$$B_\theta = \begin{cases} -A_1\left(\dfrac{\sinh(r/\lambda_L)}{r} + \dfrac{\sinh(r/\lambda_L)}{r(r/\lambda_L)^2} - \dfrac{\cosh(r/\lambda_L)}{r(r/\lambda_L)}\right)\sin\theta, & r \leqslant a \\[4mm] a_1\dfrac{\sin\theta}{r^3} - B_0\sin\theta, & r > a \end{cases} \tag{3.5}$$

Thus, figure 3.2 illustrates the field penetration process for $\lambda_L/a = 0.05$. One may notice the influence of geometry, as well as the convergence to the applied field values for the exterior zone. Recall that, being scaled by the ratio λ_L/a, penetration is also dependent on the polar angle.

Some additional comments are deserved.

- The profile of circulating currents that penetrate in the sphere is apparent just by recalling (2.14): $\mathbf{J} = -\mathbf{A}/\mu_0\lambda_L^2$
- The field created by the superconductor in the exterior region $r > a$ is exactly dipolar, independent of the value of λ_L. In particular, if one recalls the expression for the magnetic dipole

$$\mathbf{A}_{dip} = \frac{\mu_0 m_{dip}}{4\pi}\frac{\sin\theta}{r^2}\hat{\boldsymbol{\phi}}, \tag{3.6}$$

the magnetic moment of the sphere is given by

$$m_{\text{sphere}} = \frac{4\pi}{\mu_0}a_1 = \frac{2\pi B_0 a^3}{\mu_0}\left[3\frac{\lambda_L}{a}\coth\left(\frac{a}{\lambda_L}\right) - 3\left(\frac{\lambda_L}{a}\right)^2 - 1\right], \tag{3.7}$$

an expression that may be useful to evaluate limiting values of interest. For instance, for the bulk approximation $\lambda_L/a \to 0$ one recovers the well known expression for the volume magnetisation of a perfectly diamagnetic sphere

$$M_V \approx -\frac{3B_0}{2\mu_0} \tag{3.8}$$

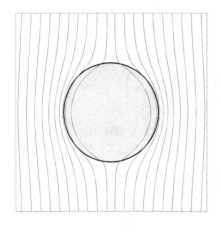 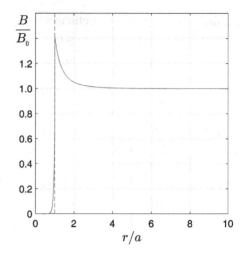

Figure 3.1. Penetration of the magnetic field in the Meissner state of a superconducting sphere of radius a for $\lambda_L = 0.05\,a$. Left: field lines. Right: field modulus $B = B_\theta(r)$ in the equatorial cross section.

Related to this, an interesting property for further discussion is the concentration of magnetic field close to the surface of the sphere that one may notice in figure 3.1. Within the bulk approximation one gets[3]

$$\mathbf{B}(r = a^+, \theta = \pi/2) = -\frac{3}{2}B_0\hat{k} \tag{3.9}$$

A very relevant consequence of the above kind of relation (3.9) is the influence of finite size effects in superconducting phenomena. Just notice that if one conducts an experiment with the applied field increasing from 0, it will be already for the value $H_a = B_0/\mu_0 = (2/3)H_c$ that the equator of the sphere will be exposed to the 'critical field' that induces the transition to the normal state. Then, some layer whose shape may be approximated by the lines in figure 3.1 will become normal. However, this will somehow release the density of lines at that location and, as a consequence, the outmost part will step backwards to superconductivity. As explained by London [3], if the applied field further increases, the process continues and the so-called **intermediate state**, consisting of a complex combination of superconducting and normal layers, will be established. A full description of the process is very complex, in particular because the laminar structure itself must be resolved. A coarse grained approximation is given in the book by London in a rather pedagogical exposition.

3.2 The resistive transition: fundamentals

The experimental verification of the zero resistance property shows that a somehow gradual transition occurs instead of a 'step-function' behaviour around T_c.

---

[3] The notation a^+ indicates that we are close to, but outside the superconductor.

This means that, in general, one may not assume a bulk transition with the whole sample jumping from the superconducting state to the normal state when T increases beyond T_c. A complete description of the involved physical processes in complex. In chapter 7 it will deserve our attention for the specific case of high-T_c superconductors. Here, we put forward the simplest possible description of the $R(T)$ curves in real conditions, based on London's model [3] for the transition of a type-I superconducting wire that transports current. The basic idea relies on the above mentioned theory of the *intermediate state*. As argued by London, geometrical effects are also a determining factor in the transition of the wire.

Recall that a long cylindrical wire of length L and radius $a \ll L$, carrying a full current I along its axis, produces an azimuthal magnetic field given by[4]

$$B_\varphi(\rho) = \frac{\mu_0 I}{2\pi\rho}; \quad \rho > a \tag{3.10}$$

regardless of the distribution $J_z(\rho)$ within the wire. In fact, in the bulk approximation $\lambda_L/a \to 0$, J_z will be basically a surface layer.

Transition induced by the current
Let us see how the intermediate state is an essential ingredient in the physics of the transition. To start with, we recall the famous *Silsbee's rule* [4]:

The 'critical current' and 'critical magnetic field' are not independent phenomena. When the current submitted is such that $B_\varphi(a) = \mu_0 H_c$, a transition to the normal state is expected. This defines the critical current

$$I_c = 2\pi a H_c \tag{3.11}$$

London's picture of the gradual transition to the normal state relies on the sketch in figure 3.2. If a current, slightly above I_c, is submitted, there will be a surface shell going normal and an interior zone of radius $r < a$ with interpenetration of normal/superconducting layers basically perpendicular to the flow. The full current will be given by the superposition

$$I = I_{\mathrm{INTERM}}(\rho < r) + I_{\mathrm{NORMAL}}(r < \rho < a) \tag{3.12}$$

Within the normal region one may straightforwardly use Ohm's law $\mathbf{J} = \sigma_n \mathbf{E}$. In fact, owing to the symmetry of the statement and considering the stationary regime, both vectors will be along the z-axis and: $\nabla \times \mathbf{E} = 0 \Rightarrow E_z = \text{constant}$. As in the coarse grained approximation this condition must hold all across the section for the averaged fields, the inner core may not be superconducting but a mixed structure as shown. Thus, for $\rho < r$ the electric field is the superposition

---

[4] As it is trivially obtained by application of Ampère's law in a cylindrical coordinate system (ρ, φ, z).

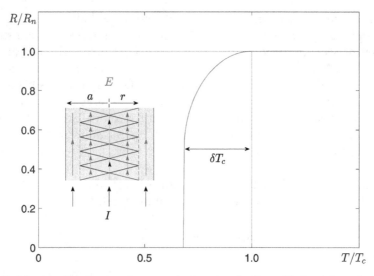

Figure 3.2. Resistive transition in a type-I superconducting wire of radius a. The inset shows the *intermediate state* structure that appears when the critical current is exceeded as predicted by F London. Blue: superconducting regions, grey: normal.

of the values $E_z = 0$ within the superconducting layers and $E_z = J_z/\sigma_n$ in between. The layers will be essentially perpendicular to the axis of the wire because of the boundary conditions for **E** and **B** at the interfaces: $E_\parallel = 0$, $B_\perp = 0$. Thus, dropping the unnecessary subindex, because all vectors are along z-axis, the above equation reads

$$I = \int_0^r J_{\text{INTERM}}(\rho)2\pi\rho d\rho + \int_r^a J_{\text{NORM}}(\rho)2\pi\rho d\rho \qquad (3.13)$$

Eventually, the wire's current–voltage law may be obtained by inserting in this equation:
  (a) Ohm's law for the normal region: $J = \sigma_n E$.
  (b) The averaged $E(J)$ law for the intermediate state region.
  (c) The value of the actual 'penetrating radius' r.

Without going into detail, here we introduce London's results

$$J_{\text{INTERM}} = \frac{H_c}{r}; \quad r = \frac{H_c}{\sigma_n E} \qquad (3.14)$$

Then, integration is simple and one gets

$$V = \frac{R_n I}{2}\left[1 + \sqrt{1 - \left(\frac{I_c}{I}\right)^2}\right], \qquad (3.15)$$

where we have used $E \equiv V/L$, and the definitions of the critical current after Silsbee (equation (3.11)) and the normal state resistance. In conclusion, the superconducting wire will be characterised by the non-linear resistive response:

$$R = \begin{cases} 0, & I < I_c \\ \dfrac{R_n}{2}\left[1 + \sqrt{1 - \left(\dfrac{I_c}{I}\right)^2}\,\right], & I > I_c \end{cases} \qquad (3.16)$$

Transition as a function of temperature

Finally, recalling the temperature dependence of the critical field, and assuming linearity close to T_c one obtains R as a function of temperature (for constant applied current I).

$$R = \begin{cases} 0, & T < T_c - \delta T_c \\ \dfrac{R_n}{2}\left[1 + \sqrt{1 - \left(\dfrac{T_c - T}{\delta T_c}\right)^2}\,\right], & T_c - \delta T_c < T < T_c \end{cases} \qquad (3.17)$$

Notice that δT_c is the interval between $R = R_n/2$ and $R = R_n$, and that it will be proportional to the measuring current I in a given experiment.

The resistive transition in a pure type-II wire involves some physical aspects different to the case considered above. One must perform the kind of analysis mentioned in section 2.6 for obtaining the equilibrium magnetisation, because the interaction between vortices (which would be annuli for the isolated wire) plays a relevant role. As emphasised by Tinkham [2], the topic is of high conceptual interest, but on the other hand, of little practical implication. Thus, recalling Silbee's rule, even for zero applied field, as soon as a current of value $2\pi a H_{c1}$ flows along the material, the inexorable flux flow phenomenon leads to dissipation. Nevertheless, for practical superconductors, i.e. those with pinning centres the physics is very different. Thus, most studies focus on the transition when it takes place under applied field and the relevant mechanism is the thermal depinning of vortices. This topic will deserve our attention in chapter 7, where specific numeric resources for analysing the problem will be supplied.

3.3 The critical state

As it was explained in section 2.6, high $-\kappa$ superconductors (extreme type-II materials) are good candidates for high field applications insofar as an effective pinning structure exists. In such a case, permanent macroscopic supercurrents may be sustained together with the corresponding high field gradients.

It is apparent that the full description of the underlying phenomena, i.e. the Abrikosov's vortex lattice plus interaction with the pinning structure, is highly involved even through the phenomenological Ginzburg–Landau (GL) theory of superconductivity. Nevertheless, many experimental facts can be explained by means of macroscopic phenomenological models which override the great mathematical difficulties of more fundamental theories [5]. This is the case of the so-called critical state model (CSM), which describes the magnetisation of strongly pinned type-II superconductors, without specifying the microscopic mechanism that controls the vortex pinning. Although, numerous modifications have improved the original idea, the essence of the model was formulated by C P Bean in the 1960s and may be found in a famous series of articles [6, 7] that have inspired a large amount of further studies.

Being of macroscopic nature, the CSM is formulated in the coarse grained 'macroscopic cell' approximation introduced in section 2.6, that relies on the presence of an enormous number of vortices in the range of interest: $H_a \gg H_{c1}$. The model is posed in terms of the fields $\mathbf{B(r)}$, $\mathbf{J(r)}$ that, in this approximation, will have the meaning of averages that smooth out the mesoscopic variations.

In its basic form, the CSM describes the magnetisation process in 1D problems, i.e. infinite slabs or cylinders under longitudinal magnetic field, for which the magnetic field has a single component, say $\mathbf{B} = B\hat{k}$. It postulates that, wherever penetrated by flux, the sample holds a steepest metastable gradient of B supported by the underlying pinning force. This is formulated via the critical state equation for the only component of the current density[5] $J = \pm J_c$, 0. Here, the so-called critical current density J_c is a phenomenological parameter related to the pinning properties of the superconductor, typically through some material-dependent relation $J_c = J_c(B)$. Then, the field penetration profile can be derived by inserting this in Ampère's law. For instance, in the slab geometry this reads

$$\frac{dB}{dx} = \pm\mu_0 J_c(B) \quad \text{or} \quad 0 \tag{3.18}$$

supplemented with the appropriate boundary conditions. With the knowledge of $B(x)$ one may calculate the volume magnetisation M_V through the relation:

$$\mathbf{M}_V = \frac{\langle \mathbf{B} \rangle}{\mu_0} - \mathbf{H}_a \tag{3.19}$$

with $\langle \mathbf{B} \rangle$ being the average of the magnetic field over the sample's volume.

Application

In order to see how the CSM is implemented in practice, we analyse a classical example, i.e. the superconducting slab subjected to an applied field parallel to the faces. Let the slab occupy the region given by $|x| \leqslant w$, $|y| \leqslant \infty$, $|z| \leqslant \infty$ and let the applied field be $\mathbf{H}_a = H_a \hat{k}$. Following Bean's work, we will assume that J_c is a constant. As shown in many textbooks, equation (3.18) together with the boundary condition $B(x = \pm w) = \mu_0 H_a$, straightforwardly lead to the field penetration profiles in figures 3.3 and 3.4, that may be generated by a straightforward 'geometrical'

---

[5] For instance $\mathbf{J} = J\hat{j}$ in slab geometry or $\mathbf{J} = J\hat{\phi}$ in cylindrical symmetry.

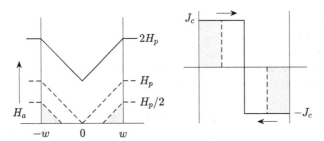

Figure 3.3. Initial magnetisation profiles for a superconducting slab in the critical state. Left: magnetic field penetration profiles $B(x)/\mu_0$. Right: evolution of the superconducting current density.

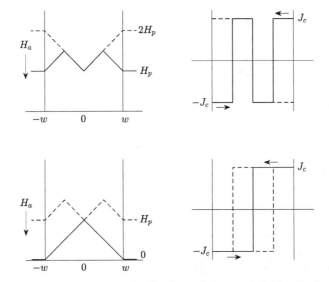

Figure 3.4. Same as figure 3.3, but for the applied magnetic field cycling back.

method[6]. Outstandingly, although implicit, there is something more behind the plots. Thus, equation (3.18) tells about the steady states with macroscopic supercurrents sustained by the existence of pinning, but the actual profiles have been determined by Faraday's law also. Just notice that consecutive profiles do not merely fulfil the critical state equation, but are just those with the smallest flux variations compatible with the boundary conditions. Although this is a trivial rule for idealised geometries, it will be an essential feature in more involved CSM problems, as it will be seen in section 8.2.

The interpretation of figures 3.3 and 3.4 is as follows. Starting from a virgin state with $B = 0$, the superconducting currents appear so as to minimise flux variations induced by the external source (ramp of H_a). When external variations cease, the eventual stationary flux penetration profile (due to the underlying pinning forces) is determined by the condition $|J_y| = J_c$.

---

[6] One must just plot parallel lines with a given slope.

An important feature of the CSM, and mostly the origin of its success, is the ability to predict the hysteretic behaviour of the magnetic response of super-conductors with strong pinning. Thus, aided by figures 3.3 and 3.4, one may straightforwardly calculate the volume magnetisation in (3.19). As shown by Bean, when the applied field is cycled, one obtains

$$\frac{M_v}{H_p} = \begin{cases} -\dfrac{H_a}{H_p} + \dfrac{1}{2}\left(\dfrac{H_a}{H_p}\right)^2; & H_a \leqslant H_p \\ -\dfrac{1}{2} & ; \quad H_a > H_p \end{cases} \tag{3.20}$$

for the initial branch of the cycle, and

$$\frac{M}{H_p} = -\frac{H_a}{H_p} + \frac{H_a H_M}{2H_p^2} \pm \frac{1}{4}\left[\left(\frac{H_a}{H_p}\right)^2 - \left(\frac{H_M}{H_p}\right)^2\right]; \quad H_M \leqslant H_p \tag{3.21}$$

for the upper/lower courses in between the cycle's limits $\pm H_M$.

In the above formulas, we have used the customary definition $H_p = J_z w$. In passing we note that such expressions follow straightforwardly by recalling that $\langle B \rangle$ is nothing but the (normalised) area under the curve $B(x)$, which can indeed be obtained by simple decomposition in polygons. The arising hysteresis cycle is shown in figure 3.5. Notice that the points of the cycle that correspond to the steps displayed in the previous plots are highlighted. Conventionally, one defines the so-called *partial penetration regime* (H_a: $0 \rightarrow H_p$) as the part of the initial ramp for which a flux free core still exists. In this simple example, the core extends over the segment $(-x_c, x_c)$ with $x_c \equiv w - H_a/J_c$ and exists until the applied field reaches the value $H_p = J_c w$, the so-called *full penetration field*. Also noticeable is the origin of hysteresis. Just comparing the two situations for which $H_a = H_p$ in this cycle, it is

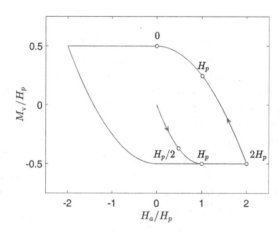

Figure 3.5. Volume magnetisation for the superconducting slab corresponding to the external field cycle: H_a: $0 \rightarrow 2H_p \rightarrow -2H_p \rightarrow 2H_p$ with $H_p = J_c w$.

apparent that the hysteretic behaviour relates to the existence of trapped flux profiles (see figure 3.4).

Some eventual remarks are due, concerning the application of the CSM in superconductivity.

- As stated, the model applies in the high-field region $H_a \gg H_{c1}$, because we have neglected the contribution of the equilibrium magnetisation curve (recall figure 2.6). Essentially, the model considers the macroscopic current density related to the gradient in the accumulation of vortices, and neglects Meissner currents. In general, this is not an important limitation in the large scale.

- The model applies in the so-called *hard superconductors*, i.e. in the limit of high pinning forces. This is a good approximation in many cases of interest (as in cuprates and MgB_2 not close to T_c), which allows to neglect such phenomena as thermally activated relaxation or the annihilation of contiguous parallel antivortices[7].

- The approximation $J_c \approx$ constant, introduced by Bean, strongly simplifies calculations and allows to obtain qualitative or semi-quantitative agreement between theory and experiment in many cases. Nevertheless, if one needs to go quantitative, the model must incorporate some realistic[8] dependence of the critical current density on the local magnetic field $J_c(B)$. Since the pioneering work by Kim and co-workers [8], who thoroughly considered the topic and issued their celebrated model, there have been lots of investigations on this. Many theoretical and experimental works describe methods to derive the correct $J_c(B)$ dependence for each material (as explained in section 9.3.3). In our case, being concentrated on the numerical solutions, we will provide a basic algorithm that the reader may modify according to the specific case.

- Apparently, the basic critical state equation (3.18) is readily applicable when there is a unique, known in advance, direction for the current and the field depends on one independent variable, as in the mentioned idealised geometries. However, this does not suffice to solve the seemingly simple situation in which the external field is applied perpendicular to the axis of a cylindrical sample (unless further assumptions are made). In such case and many other situations related to practical applications, the magnetic field will have at least a 2D character and the model must be elaborated further. In fact, multicomponent problems are in most cases to be solved by numerical procedures. A good part of this book (chapters 8 and 9 in particular) will be devoted to providing the tools for such problems.

- From the physical point of view, Bean's approximation has to be interpreted as sketched in figure 3.6. The evolution between one critical state and another is induced by the appearance of electric fields due to incoming/outcoming flux triggered by some external source. Thus, when the maximum field gradient is

---

[7] Chapter 7 is devoted to the influence of thermal activation close to T_c. Other relaxation effects are also treated in sections 8.5 and 8.6, so as to enable comparison between more and less *hard* materials.

[8] Strictly speaking, in fact it is the pinning force i.e. $J_c \cdot B$ (recall (2.46)) that should be more reasonably bounded by a constant.

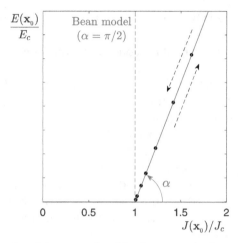

Figure 3.6. Model approximation of the process describing flux penetration in a superconductor with pinning. At a given point x_0, an electric field is induced if magnetic flux changes, and a transient overcritical current appears. When variations cease, J goes to the equilibrium value J_c and E to 0. The behaviour for $J < 0$ may be obtained by symmetry: $E(J < J_c) = \rho_f(J + J_c)$.

exceeded, and the pinning force threshold ($J > J_c$), dissipation takes place. At a first approximation, for those points of the sample where this occurs, electric fields ramp up and down along the line $\rho_f(J - J_c)$ as indicated in figure 3.6. The limiting case $\alpha \to \pi/2$, i.e. very high resistivity, corresponds to the CSM approximation, for which a nearly instantaneous response is expected due to the boundless dissipation[9]. Thus, one assumes that critical profiles corresponding to the modified boundary conditions appear without delay. For AC processes this means that the evolution of the response is exactly given by the frequency of the applied excitation. In chapter 8, we will deal with methods that allow to introduce finite values of ρ_f.

3.4 Josephson junctions

As stated in the previous chapter, since the early 1950s the understanding of the quantum nature of superconductivity progressively lead to fundamental advances in the field: (i) London's prediction of flux quantisation and its later observation and (ii) the formulation of the Ginzburg–Landau theory, with such transcendental applications as the description of type-II superconductivity by A A Abrikosov. In the early 1960s, a new pathbreaking phenomenon came to the scene. As predicted by B D Josephson [9, 10], owing to the quantum behaviour of charge carriers in superconductors, tunnelling currents between neighbouring superconductors may occur.

A number of striking features that we will describe below in some detail were foreseen by Josephson. In the junction between two superconductors: (i) a DC

---

[9] Recall that the typical relaxation time for a linear resistive process is given by $\mu_0 \ell^2/\rho$ with ℓ a characteristic length of the sample.

supercurrent may occur at zero voltage with a maximum value, say I_c, (ii) AC oscillating currents occur at finite DC voltages, and (iii) the maximum supercurrent may be modulated by the applied magnetic field, i.e. one has $I_c(B)$. This amazing behaviour has enabled applications in a variety of fields, such as electronics, metrology and recently in the challenging area of quantum computing.

3.4.1 Josephson relations

The fundamental equations that describe the phenomenon (Josephson relations) may be obtained by application of the wave function formulation introduced in section 2.4. As sketched in figure 3.7, below we will show how these relations emerge in the junction between two massive superconducting electrodes connected by a tiny quasi-1D bridge. As clarified by Tinkham [2], although Josephson made his prediction for the case of two superconductors separated by a barrier, it is nowadays clear that the same physics arises for very general 'weak links' between super-conductors[10]. Here, assuming that the electrodes and the bridge are all made of the same material we will proceed with a wave function expression written as $\Psi = \Psi_0 \exp[i\theta(\mathbf{r}, t)]$. The equations describing (i) the *zero voltage supercurrent* and (ii) the time dependent *voltage–phase relation* are sequentially analysed.

Zero voltage relation
The appearance of the superconducting tunnelling current may be formulated on the basis of the first GL equation (see equation (2.33)). Let us work it out in two steps. First, we will use the simplifying hypothesis that $\mathbf{A} = 0$. Then one has

$$\alpha\Psi + \beta|\Psi|^2\Psi - \frac{\hbar}{2m}\Delta\Psi = 0 \qquad (3.22)$$

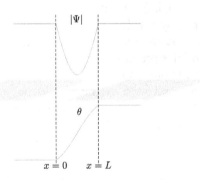

Figure 3.7. Josephson junction defined by a thin bridge between two massive superconducting electrodes. The modulus and phase along the junction for $\Delta\theta = \pi/2$ are shown (by plotting equation (3.25)).

---

[10] For a pedagogical derivation of Josephson relations in a more general context the reader is addressed to the book by de Gennes [11].

Now, assuming that the bridge extends from $x = 0$ to $x = L$ and that the third term of the equality is predominant[11], the physics of the problem is obtained by solving Laplace's equation

$$\Delta \Psi = 0, \qquad (0 < x < L) \tag{3.23}$$

with the boundary conditions

$$\begin{aligned} \Psi(x \leqslant 0) &= \Psi_0 \\ \Psi(x \geqslant L) &= \Psi_0 e^{i\Delta\theta} \end{aligned} \tag{3.24}$$

Physically, this means that although identical, the superconducting electrodes in equilibrium have a common density of superelectrons ($|\Psi_0|^2$), but may hold a difference in phase ($\Delta\theta$).

Apparently, under the above conditions, the complex order parameter along the bridge obeys the equation

$$\Psi(x) = \Psi_0\left(1 - \frac{x}{L} + \frac{x}{L}e^{i\Delta\theta}\right) \equiv |\Psi|(x)\, e^{i\theta(x)} \tag{3.25}$$

Then, the supercurrent density is given by

$$\mathbf{J} = \frac{q\,\hbar}{2\,m\,i}(\Psi^*\nabla\Psi - \Psi\nabla\Psi^*) = \frac{q\,\hbar\,\Psi_0^2}{mL}\sin(\Delta\theta)\,\hat{\imath} \tag{3.26}$$

By using S for the cross sectional area of the bridge, one gets the full current (notice the orientation in terms of the signs of q and $\Delta\theta$)

$$I = \frac{2e\,\hbar\,\Psi_0^2 S}{mL}\sin(\Delta\theta) \equiv I_c \sin(\Delta\theta) \tag{3.27}$$

Interestingly, this equation indicates that a DC supercurrent may be maintained by a phase gradient (and not by a voltage drop as in conventional conductors).

Let us now discuss the modification required in the previous equation when a non-zero vector potential is present. We will rely on the concept of gauge invariance. In fact, any acceptable expression of a physical quantity should be invariant under gauge transformations, and as it is this would not be the case for I. However, recalling section 2.4.2 invariance for the current density was ensured by the relation $\chi = (2\pi/\Phi_0)\theta$, with χ the gauge field for the vector potential. Then, as one should obtain the same physics using $\mathbf{A} = 0$ and $\mathbf{A}' = \nabla\chi$, the phase should transform according to $\Delta\theta \to \Delta\theta' = \Delta\theta - (2\pi/\Phi_0)\Delta\chi$. As a consequence, Josephson's relation becomes invariant if one uses the so-called *gauge invariant phase difference*:

$$\Delta\varphi = \Delta\theta - \frac{2\pi}{\Phi_0}\int_{x=0}^{x=L} \mathbf{A} \cdot d\ell \tag{3.28}$$

---

[11] This is justified for short bridges [2] whose length is small as compared to the typical distance over which Ψ changes in a bulk. Mostly, variations take place along the bridge.

with $\Delta\theta$ to be understood as $\theta(x = L) - \theta(x = 0)$.

Eventually, the *zero voltage relation* takes the form

$$I = I_c \sin(\Delta\varphi) \tag{3.29}$$

The voltage–phase relation

Next, we will analyse the situation that relates to the appearance of AC oscillations reported above. One may proceed by taking the time derivative of equation (3.28) and recalling the Scrödinger-like equation for the evolution of the superconducting order parameter (equation (2.28)). After some algebra this leads to

$$\frac{\partial(\Delta\varphi)}{\partial t} = -\frac{\Lambda_L}{2\Psi_0^2 \hbar}(J^2(L) - J^2(0)) - \frac{2\pi}{\Phi_0}(\phi(L) - \phi(0)) - \frac{2\pi}{\Phi_0}\int_0^L \dot{\mathbf{A}} \cdot d\ell \tag{3.30}$$

which simplifies to[12]

$$\frac{\partial(\Delta\varphi)}{\partial t} = \frac{2\pi}{\Phi_0}\int_0^L \mathbf{E} \cdot d\ell = \frac{2\pi}{\Phi_0}V \tag{3.31}$$

As the right-hand-side of this equation is nothing but the voltage drop across the junction[13], this is the celebrated *voltage–phase relation*. Equations (3.29) and (3.31) are the so-called *Josephson relations* and will be used to describe quantitatively the main features of the phenomenon. Thus

(i) Equation (3.29) describes the DC supercurrent across the junction and predicts a maximum value I_c related to microscopic parameters of the superconductor, as well as to fundamental constants.

(ii) If a DC voltage V_{DC} is applied to the junction, equation (3.31) predicts a phase variation linear in time, that translated to equation (3.29) implies the appearance of an AC current with a characteristic time constant given by Φ_0/V_{DC}.

Concerning the modulation of the maximum supercurrent by an applied magnetic field $I_c(B)$, it will be analysed in the forthcoming sections by a more elaborate analysis of equation (3.29). First, in section 3.4.2 we develop the concept of SQUID magnetometry, based on the phenomenon of flux quantisation within a superconducting loop interrupted by a pair of point-like Josephson junctions. Next, in section 3.4.3, we concentrate on the effect of flux quantisation in a single extended Josephson junction. As the reader may promptly notice, there is a tight analogy

---

[12] By recalling the MQS continuity condition $J(x = 0) = J(x = L)$ and the expression for the electric field $\mathbf{E} = -\nabla\phi - \partial\mathbf{A}/\partial t$.

[13] We call the readers' attention that the current density flows from $x = 0$ towards $x = L$ and the calculated voltage drop is $V \equiv V(0) - V(L)$.

between these phenomena and those behind Young's double slit experiment and Fraunhofer's diffraction in optics.

3.4.2 The SQUID magnetometer

The inset of figure 3.8 shows the basic structure of a Superconducting QUantum Interference Device (SQUID), i.e. a loop of superconducting material, interrupted by a pair of junctions, biased by a current, and exposed to perpendicular magnetic flux Φ_{ext}. This device was proposed by Jaklevik and co-workers in 1964 [12] soon after Josephson's work, and has ever been a key application of superconductivity for its ability to detect the tiniest magnetic fields. In order to understand how it works, below we analyse the fundamental physical principles[14]. To start with, let us consider the sketch in figure 3.8. In particular, being interested in the modulation of the critical current by the magnetic flux, i.e. $I_c(B)$ we concentrate on the evaluation of phase variations $\Delta\varphi$ in order to apply Josephson's zero voltage relation (3.29). For such purpose, we state the problem as follows[15]

- The two junctions are identical ($I_{c1} = I_{c2} \equiv I_c$) and of negligible size.
- The magnetic flux is given by the action of some external source (i.e. induction effects of the loop are negligible): $\Phi_{\text{ext}} = B_{\text{ext}} S_{\text{loop}}$.
- The superconducting tracks may be considered ideal (flux penetration effects are negligible).

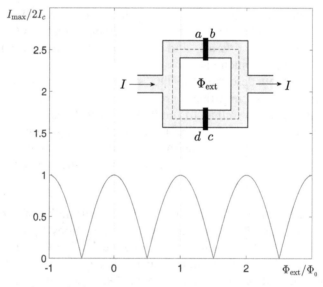

Figure 3.8. Sketch of a superconducting quantum interference device (SQUID). Below, we plot the maximum supercurrent through the device in terms of the normalised magnetic flux across the area of the loop.

---

[14] A vast amount of literature is at hand to go deeper in this subject, both as a part of general superconductivity textbooks [2, 13], in more specialised books [14] or even dedicated monographs [15, 16].

[15] A number of simplifications are made in order to extract the basic physics with the lesser complication.

Recall that, under the above conditions, the integration of the vector potential along the indicated circuit relates to the quantity of interest Φ_{ext}

$$\oint \mathbf{A} \cdot d\ell = \Phi_{\text{ext}} \tag{3.32}$$

Now, making use of relation (3.28), this may be related to the changes of phase through the junctions. For this purpose, we detach the line integration as follows:

$$\oint \mathbf{A} \cdot d\ell = \int_a^b \mathbf{A} \cdot d\ell + \int_b^c \mathbf{A} \cdot d\ell + \int_c^d \mathbf{A} \cdot d\ell + \int_d^a \mathbf{A} \cdot d\ell \tag{3.33}$$

Then, recalling that within the superconductor one has $\Lambda_{L} \mathbf{J} = - (\Phi_0/2\pi)\nabla\theta - \mathbf{A}$, considering $J \approx 0$, and applying (3.28) we can write

$$\oint \mathbf{A} \cdot d\ell = - \frac{\Phi_0}{2\pi}\left[\Delta\theta_a^b - \Delta\varphi_a^b + \Delta\theta_b^c + \Delta\theta_c^d - \Delta\varphi_c^d + \Delta\theta_d^a\right] \tag{3.34}$$

Considering that $\Delta\theta_a^b + \Delta\theta_b^c + \Delta\theta_c^d + \Delta\theta_d^a$ evaluates the variation of the order parameter's phase in a close path, a quantity that is defined 'modulo' 2π, one eventually obtains

$$\frac{-\Phi_0}{2\pi}\left(-\Delta\varphi_a^b - \Delta\varphi_c^d + 2n\pi\right) = \Phi_{\text{ext}} \tag{3.35}$$

that is to say

$$\Delta\varphi_a^b + \Delta\varphi_c^d = \frac{2\pi\Phi_{\text{ext}}}{\Phi_0} + 2n\pi \tag{3.36}$$

Eventually, one may apply (3.29) and use Kirchhoff's law to describe the bifurcation of current in the loop:

$$I = I_c \sin(\Delta\varphi_a^b) + I_c \sin(-\Delta\varphi_c^d), \tag{3.37}$$

where the minus sign comes from the orientation of our original circuit for the line integration (opposite to the expression of Kirchhoff's law). By applying elementary trigonometric relations, one gets

$$I = 2I_c \cos\left(\pi\frac{\Phi_{\text{ext}}}{\Phi_0}\right) \sin(\Delta\varphi_a^b + \pi\frac{\Phi_{\text{ext}}}{\Phi_0}) \tag{3.38}$$

For a a given value of the external magnetic flux, this expression is maximised when the phase $\Delta\varphi_a^b$ is such that the 'sin()' term evaluates to ± 1. Thus, the maximum supercurrent allowed through the system is

$$I_{\text{max}} = 2I_c \left| \cos\left(\pi\frac{\Phi_{\text{ext}}}{\Phi_0}\right) \right| \tag{3.39}$$

This dependence is displayed in figure 3.8, and is the keypoint for the detection of magnetic fields. Notice that the current is a periodic function in terms of the flux quantum Φ_0. In order to make an estimate of sensitivity in the detection of magnetic fields, let us assume an area of the loop about 1 cm^2. This gives a basic sensitivity (period of the critical current signal) of $\approx 2 \times 10^{-11}$ T.

As the reader may recall, the above exposition may oversimplify reality. Certainly, the practical facts are more complex, and a number of technical details arise when a real field detector is built and operated. In particular, the so-called dc-SQUID, that is well described by the above equations is typically operated beyond the critical current value, i.e. at finite voltage. With this, several effects as noise and hysteresis effects reduction are aimed [2]. In order to provide a minimal illustration of this fact, we introduce figure 3.9. As it will be shown in chapter 15 of this book, by carefully analysing equations (3.29) and (3.31) one may demonstrate that when an overcritical current is submitted through a junction (with critical current $I_{\max}$) in parallel with a shunt resistance R, an oscillating voltage appears, with an average value given by

$$\langle V \rangle = IR \sqrt{1 - \left(\frac{I_{\max}}{I}\right)^2} \tag{3.40}$$

Applied to the case under consideration (diagram in figure 3.9), this gives

$$\langle V \rangle = I \frac{R}{2} \sqrt{1 - \left[\frac{2I_c}{I} \cos\left(\frac{\pi \Phi_{\text{ext}}}{\Phi_0}\right)\right]^2} \tag{3.41}$$

Here, the respective factors 1/2 and 2 appear as related to the equivalence for either two resistances in parallel $R_{\text{equiv}} \to R/2$ or two junctions in parallel ($I_{c,\text{equiv}} \to 2I_c$, i.e. each junction carries its maximum).

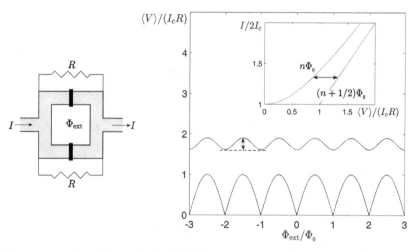

Figure 3.9. Response of a resistively shunted SQUID magnetometer to the piercing external magnetic flux. The oscillatory behaviour of the average voltage across the device is shown for two overcritical values of the bias current ($I = 2I_c$ (lower curve) and $I = 3.8I_c$ (upper curve)). The inset displays the $I(\langle V \rangle)$ behaviour for $\Phi_{\text{ext}} = n\Phi_0$ and $\Phi_{\text{ext}} = (n + 1/2)\Phi_0$.

The above dependence is shown in figure 3.9. $\langle V \rangle$ is plotted as a function of the applied flux Φ_{ext} for two different values of the bias current ($I = 2I_c$ and $I = 1.9 \times 2I_c$). We notice that the largest contrast in voltage appears for $I = 2I_c$ with the voltage ranging from 0 (for $\Phi_{ext} = n\Phi_0$) to I_cR (for $\Phi_{ext} = (2n + 1)/2\Phi_0$). This important property is emphasised in the inset, by the indication of the distance between the $I(\langle V \rangle)$ curves for the 'extreme' values of the magnetic flux (Φ_{ext}). As a consequence, this gives a rule of operation. Optimally, one polarises the SQUID to the bias $I = 2I_c$, and quantifies the external flux in units of Φ_0 by 'counting' the number of jumps in the voltage signal as the external flux increases from 0 to Φ_{ext}.[16]

3.4.3 Josephson effect in extended junctions

Let us now analyse the influence of the applied magnetic field on the maximum current that flows through a Josephson junction that may not be considered point-like as in the previous section. In this context, phase variations are not only relevant 'across' the junction, but also 'along' its extension (say $\Delta\varphi(y)$ in figure 3.10). In this chapter, nonetheless, we still keep at the level that allows to assume that self-field effects may be neglected, so that the magnetic flux is constant and given by the external action. An application to the more realistic situation that accounts for such effects will be discussed in section 15.1.

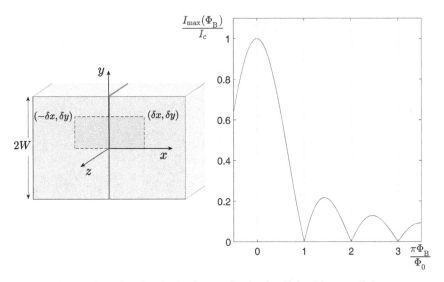

Figure 3.10. (Left) Josephson junction in the form a flat barrier (defined by $x = 0$) between two superconducting blocks. The shaded rectangle (XY plane) is defined so as to perform line integration along its (dashed) perimeter. To the right, we show the (symmetric) modulation of the tunnelling current across the junction due to the flux of $\mathbf{B} = B_0\hat{k}$.

---

[16] The actual protocol to 'increase' the applied field may be realised in practice by some active process of the instrument. For instance, one may apply an extra field created by the instrument, that is opposite to the original, and is ramped until the original is nulled.

Let us consider the model depicted in figure 3.10. Two superconducting blocks are separated by a flat barrier, while a uniform magnetic field of induction B_0 is applied along the z-axis. We will use equation (3.29) in order to evaluate such maximum current, but now considering that $\Delta\varphi$ may vary not only across, but also along the barrier, i.e. the phase difference between the banks may be a function of position $\Delta\varphi(y)$.

We start by the statement of equation (3.28) applied to the vertices of the rectangle highlighted in the figure. Notice that, evaluated within each superconducting block, this relation gives

$$\varphi(\delta x, \delta y) - \varphi(\delta x, 0) = \theta(\delta x, \delta y) - \theta(\delta x, 0) - \frac{2\pi}{\Phi_0} \int_{(\delta x,\, 0)}^{(\delta x, \delta y)} A_y \, dy$$

$$\varphi(-\delta x, \delta y) - \varphi(-\delta x, 0) = \theta(-\delta x, \delta y) - \theta(-\delta x, 0) - \frac{2\pi}{\Phi_0} \int_{(-\delta x,\, 0)}^{(-\delta x, \delta y)} A_y \, dy$$

$$(3.42)$$

Subtracting, and using $\Delta\varphi(\delta y) \equiv \varphi(\delta x, \delta y) - \varphi(-\delta x, \delta y)$ and so on, we get

$$\Delta\varphi(\delta y) - \Delta\varphi(0) =$$

$$\Delta\theta(\delta y) - \Delta\theta(0) - \frac{2\pi}{\Phi_0}\left(\int_{(\delta x,\, 0)}^{(\delta x,\delta y)} A_y \, dy - \int_{(-\delta x,\, 0)}^{(-\delta x,\delta y)} A_y \, dy \right) \qquad (3.43)$$

Next, we notice that $\Delta\theta(\delta y)$ and $\Delta\theta(0)$ may be written in terms of $\mathbf{A}$. Thus, within each superconducting block, one may use equations (2.25) and (2.27), so as to obtain

$$\Lambda_L \mathbf{J} = - \frac{\Phi_0}{2\pi}\nabla\theta - \mathbf{A} \qquad (3.44)$$

In particular this gives

$$\Delta\theta = \int \nabla\theta \cdot d\ell = - \frac{2\pi}{\Phi_0} \int \Lambda_L \mathbf{J} \cdot d\ell + \frac{2\pi}{\Phi_0} \int \mathbf{A} \cdot d\ell \qquad (3.45)$$

When this is applied to the horizontal segments ($[-\delta x, \delta y] \rightarrow [\delta x, \delta y]$ and $[-\delta x, 0] \rightarrow [\delta x, 0]$) one gets

$$\Delta\theta(\delta y) = \frac{2\pi}{\Phi_0} \int_{(-\delta x,\, \delta y)}^{(\delta x,\delta y)} A_x \, dx$$

$$\Delta\theta(0) = \frac{2\pi}{\Phi_0} \int_{(-\delta x,\, 0)}^{(\delta x, 0)} A_x \, dx$$

$$(3.46)$$

as long as (i) either $J = 0$ well within the electrode or (ii) $\mathbf{J}\perp d\ell$ close to the barrier[17].

---

[17] Physically, the approximation leading to the fulfilment of such conditions corresponds to a geometry of contact for which the length along y-axis (i.e. along the junction itself), say $2W$ is dominant. The longer the dimension along y, the bigger the region around the centre of the superconductor for which the screening currents run vertical and concentrate close to the boundary. See figure 15.1 and the accompanying explanations, with the exact solution of the problem.

Eventually, one has the result

$$\Delta\varphi(\delta y) - \Delta\varphi(0) = \frac{2\pi}{\Phi_0} \oint \mathbf{A} \cdot d\ell \tag{3.47}$$

where the closed path integration is done along the (dashed) perimeter of the shaded rectangle in figure 3.10.

Now, recalling that the path integration of the vector potential is nothing but the magnetic flux across the enclosed section

$$\Delta\varphi(\delta y) - \Delta\varphi(0) = \frac{2\pi}{\Phi_0} B_0 \delta x \, \delta y \tag{3.48}$$

which implies

$$\frac{\partial \Delta\varphi}{\partial y} = \frac{2\pi}{\Phi_0} B_0 \delta x$$

$$\Downarrow$$

$$\Delta\varphi(y) = \left(\frac{2\pi}{\Phi_0} B_0 \delta x\right) y + \Delta\varphi(0)$$

Now, having obtained $\Delta\varphi(y)$ we are ready to evaluate the effect of **B** on the current through the junction. Taking into account a non-uniform current density

$$J(y) = J_c \sin[\Delta\varphi(y)] \tag{3.49}$$

one may obtain the full current by integration along the junction, i.e.

$$I = \iint J(y)dydz = I_c \frac{\sin(\pi\Phi_B/\Phi_0)}{\pi\Phi_B/\Phi_0} \sin[\Delta\varphi(0)] \tag{3.50}$$

with Φ_B being the flux defined by $B_0 \delta x(2W)$ and customarily named after *flux through the junction*.

Equation (3.50) may be understood as the current through the junction for a given value of the applied field B_0 and a given phase difference in the central part of the superconductor $\Delta\varphi(0)$. It is obvious that the maximum current allowed for a given value of the magnetic field will occur when the phase difference is established to be such that $\sin[\Delta\varphi(0)] = \pm 1$. Then, the expression for the maximum allowed current is

$$I_{max} = I_c \left| \frac{\sin(\pi\Phi_B/\Phi_0)}{\pi\Phi_B/\Phi_0} \right| \tag{3.51}$$

This is the celebrated 'Fraunhofer pattern' for the maximum current through the junction. It is depicted in figure 3.10.

Further considerations related to this topic may be found in the literature. Of particular pedagogical value is the exposition by T P Orlando [13]. On the other hand, some issues of practical interest related to realistic effects in extended

Josephson junctions are touched in chapter 15 of this book, where the reader is also provided with related software utilities.

References

[1] Meissner W and Ochsenfeld R 1933 Ein neuer effekt bei eintritt der supraleitfähigkeit *Naturwissenschaften* **21** 787–8

[2] Tinkham M 1996 International Series in Pure and Applied Physics *Introduction to Superconductivity* (New York: McGraw Hill)

[3] London F 1950 *Superfluids* vol 1 (New York: Wiley)

[4] Silsbee F B 1917 Note on Electrical Conduction in Metals at Low Temperatures *Bureau of Standards Scientific Paper No. 307*

[5] Campbell A M and Evetts J 1972 Monographs on Physics: Taylor and Francis *Critical Currents in Superconductors* (London: Taylor & Francis)

[6] Bean C P 1962 Magnetization of hard superconductors *Phys. Rev. Lett.* **8** 250–3

[7] Bean C P 1964 Magnetization of high-field superconductors *Rev. Mod. Phys.* **36** 31–9

[8] Kim Y B, Hempstead C F and Strnad A R 1963 Magnetization and critical supercurrents *Phys. Rev.* **129** 528–35

[9] Josephson B D 1962 Possible new effects in superconductive tunnelling *Phys. Lett.* **1** 251–3

[10] Josephson B D 1965 Supercurrents through barriers *Adv. Phys.* **14** 419–51

[11] de Gennes P G 1999 *Superconductivity of Metals and Alloys* (Cambridge, MA: Perseus)

[12] Jaklevic R C, Lambe J, Silver A H and Mercereau J E 1964 Quantum interference effects in Josephson tunneling *Phys. Rev. Lett.* **12** 159–60

[13] Orlando T P and Delin K A 1991 *Foundations of Applied Superconductivity* (Reading, MA: Addison-Wesley)

[14] Barone A and Paternò G 1982 *Physics and applications of the Josephson effect* (New York: Wiley)

[15] Clarke J and Braginski A I 2006 *The SQUID Handbook I: Fundamentals and Technology of SQUIDs and SQUID Systems* vol 1 (New York: Wiley)

[16] Clarke J and Braginski A I 2006 *The SQUID Handbook II: Applications of SQUIDs and SQUID Systems* vol 2 (New York: Wiley)

Chapter 4

Some revealing experiments with superconductors

This chapter aims at an elementary introduction to some selected experimental techniques that reveal the basic properties of superconducting materials. It is not our purpose to reach the level of detail necessary to perform actual quantitative experiments. For such purpose, the reader may consult the suggested references, where exhaustive explanations on the experimental details are offered. Mainly, our purpose is to provide a link between the theoretical concepts and the real facts in the lab, i.e. what is measured and how.

The chapter is organised on the basis of the following categories:

- The transition to zero resistance as straightforwardly quantified by *transport measurements*.
- The penetration and exit of magnetic field in different types of superconductors, as evaluated by *inductive measurements*.
- The formation of macroscopic current density paths, responsible for the magnetic flux penetration or exit, straightly visualised by *magneto-optical* techniques.
- The appearance of electromechanical forces between magnets and superconductors, as directly quantified both for large scale applications and related to the evaluation of London's penetration depth in the microscale.

4.1 Transport measurements

The verification of the zero resistance property has been a predominant technique in the research of superconducting materials. First, it is the obvious physical characterisation of any sample. As a preliminary step of any investigation, one must verify the transition to the superconducting state at low temperatures. On the other hand, by going a step further, if the transition is studied under applied magnetic field, pressure, while rotating the sample in field, etc a big amount of additional information

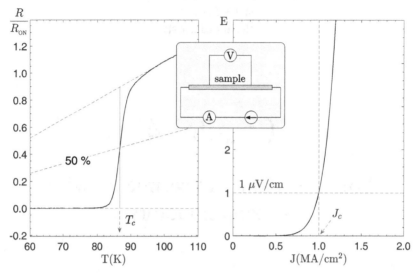

Figure 4.1. Sketch of conventional resistive measurements. Left: typical resistance versus temperature curve for YBCO films at zero applied field. Right: typical isothermal voltage–current characteristic converted to electric field versus current density.

concerning the superconducting mechanisms may be obtained. Some aspects and tools for the analysis of $R(\mathbf{B},T)$ measurements will be treated in chapter 7.

Below, we describe the basic experimental scheme conventionally used in resistive measurements. As shown in figure 4.1, the underlying principle is rather simple. One needs a controlled current source that delivers current to the sample, and a voltmeter that detects the voltage drop. In order to cancel out possible voltage drops in the contacts between wires and sample, a 4-probe technique is conventionally used. As shown in the plot, one injects current through the ends and records voltage in the central part of the superconductor. Obviously, the sample is situated in a cryogenic system (for example the one in figure 4.2), allowing to change temperature along the measurement. Depending on the exact purpose of the experiment[1], the degree of sophistication of the experimental setup changes a lot, but the final operation is as said above. Experiments may be classified in two main groups as shown in figure 4.1, that illustrates typical measurements in yttrium barium copper oxide (YBCO) films.

Resistivity measurements

In this case, one records the voltage drop for a 'small enough' current through the sample, as a function of temperature. In practice, small enough means that one reaches a level for which the resistance is basically independent of the applied current (recall section 3.2). This reproduces the pioneering Kamerlingh Onnes' experiment [1–3]. As shown in the left part of the plot, cooling down the sample gives way to the superconducting transition for the YBCO superconductor. In a

---

[1] Actual range of temperature, applied magnetic field, required level of current, etc.

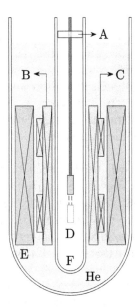

Figure 4.2. Sketch of a conventional magnetometer. A: vacuum closure, B: primary coil, C: secondary coils, D: sample, E: superconducting magnet, F: experimental space.

similar fashion to the example provided in section 3.2 for type-I materials, it is apparent that the temperature that characterises the transition, T_c (a fundamental parameter in materials for applications) must be determined by following some convention. In fact, the actual $R(T)$ curves for high-T_c superconductors show a noticeably broadened transition. An extended technique (as indicated in the plot) consists of linear fitting the normal state resistance, plotting the corresponding line multiplied by 0.5, and intersecting the result with the actual $R(T)$ [4]. For the case in figure 4.1, we show that this procedure is consistent with another popular method that defines the critical temperature through the position of the maximum of the curve dR/dT, this is marked by a vertical line in the plot.

Critical current measurements

In this case, one works under isothermal conditions, and ramps the applied current while recording voltage. Then, a more or less sharp transition occurs when a certain level of current I_c is reached. A widely accepted characterisation criterion is as follows. One converts the voltage drop to electric field dividing by the distance between the voltage pads: $E \equiv V/d$. Also, the measured current is divided by the sample's cross sectional area, thus defining the current density $J \equiv I/A$. Eventually, the characteristic critical current density is identified as that value for which the electric field reaches a conventional threshold value of 1μV/cm.

4.2 Inductive measurements

The experiment of Meissner and Ochsenfeld [5] triggered an ever extensive experimental activity, so as to record the magnetic properties of superconductors.

Measurements of the samples' magnetisation as a function of the applied magnetic field, temperature and under different cooling conditions have provided a large amount of information that has guided the search and optimisation of new materials, as well as their possible implementation in different devices. In parallel, the development of theoretical models that allow understanding of the nature of the underlying phenomena have evolved in synergy. For an extensive guide to the relevance of magnetic measurements in the development of superconductivity (mainly in the high-T_c era) the reader is addressed to the book by Hein, Francavilla and Liebenberg [6].

The basic experiment, that will be described below, relies on a rather simple operation. In a coil system, one moves the sample from the centre of one secondary coil to another in the presence of a magnetic field, and records the induced electromotive force. The result will be proportional to the sample's magnetic moment $\boldsymbol{\mu}$. In order to probe the superconducting properties, the sample is placed in the so-called *experimental space* (see figure 4.2), where one may change and record temperature, and apply a magnetic field, which typically ranges to several Tesla. As shown in the figure, this requires some cryogenic equipment. On the one hand, the superconducting material must be submitted to low temperatures in the experimental area (recall, for instance that for YBCO, $T_c \approx 90\mathrm{K}$). On the other hand, the high field is produced by means of superconducting solenoids that must be immersed in liquid He (4.2K). Thus one needs a system of cryostats as shown.

Apparently, with the 'basic' instrument shown in the figure, one may apply a full magnetic field along the axis of the system, that combines the action of the superconducting solenoid, and the primary coil. Customarily this is indicated by $B_a\hat{\boldsymbol{k}} + b_a\hat{\boldsymbol{k}}$, so as to emphasise that $B_a \gg b_a$. Also customarily, the primary may work both in dc and ac mode, i.e. $b_{ac}(t) \equiv b_a \cos(\omega t)$.[2]

Basic experiment: quantitative analysis

In a typical experiment, the sample is moved between the central position of two identical secondary coils connected in series opposition ('extracted' from one and 'introduced' into the other)[3]. All along the path, the applied magnetic field is assumed to be highly uniform. The induced electromotive force arises as follows. At first, the sample is basically linked to the first secondary coil, say by an amount of flux Φ_1. In the second position, the same occurs, but now with the secondary coil, say by Φ_2. Owing to the configuration of series opposition, and assuming a perfectly compensated system $\Phi_2 = -\Phi_1$. The total flux increment along the path of the sample is

$$\Delta\Phi = -2\Phi_1 \tag{4.1}$$

---

[2] Some pathbreaking results about the physics of vortices [7] have been obtained by using experimental setups that allow to use perpendicular field configurations, i.e. $\mathbf{B}_a \perp \mathbf{b}_{ac}$.

[3] This configuration offers some advantages as the compensation of background electromotive forces in *ac* operation.

Then, as this flux will be proportional to the sample's magnetic moment by some geometrical coefficient: $\Phi_1 = C(x_0, y_0, z_0) \, m_z$, with C being determined by the coils geometry[4], we have

$$-\Delta\Phi = \int_0^{\Delta t} \mathcal{E}(t)dt = 2 \, C \, m_z \qquad (4.2)$$

According to this, one may obtain the sample's magnetic moment by recording and integrating the electromotive force, and using C as a calibration constant that may be obtained through a measurement of some reference sample in the same instrument.

Interpretation of results

Magnetic measurements have been routinely used to obtain superconducting parameters related to the dynamics of flux penetration and exit, vortex dynamics, etc. Here, being concentrated on the macroscopic phenomena, we will comment on the critical current density parameter J_c (see section 3.3) which determines the performance of the material in practical applications. Figure 4.3 shows typical results for the hysteresis loops obtained for a type-II superconductor in isothermal conditions under a cycle of high applied magnetic field. Experimentally, this is performed by ramping the current applied to the superconducting coil in figure 4.2. Some relevant details must be taken into consideration.

- Depending on the material and on the temperature, one may obtain a behaviour that either matches the left part or the right part of figure 4.3. The former case corresponds to a field-independent J_c, while the latter reflects a relevant field-dependence $J_c(B)$. The basic interpretation may be performed in view of equations (3.21) or (9.10), that apply to the slab or cylindrical symmetries. For the latter, it is apparent that the sample's volume magnetisation saturates to the value

$$M_p = \frac{J_c R}{3} \qquad (4.3)$$

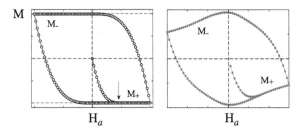

Figure 4.3. Magnetic moment per unit volume obtained for type-II superconductors at low temperatures.

---

[4] Here, the size of the sample is assumed to be negligible as compared to the dimensions of the detection system.

with R being the radius of the cylinder. Thus, by defining the volume magnetisation $M_{VOL} \equiv m_z/\text{vol}$ and by inversion of (4.3), one may obtain the sample's critical current density. Clearly, one must be sure that the magnetisation has reached saturation. Following Campbell and Evetts [8], this basic idea may be upgraded to account for several relevant effects. Additional 'offset' magnetisation due to superconducting currents of nature other than the critical state profiles[5] may be 'compensated' by subtracting the field decreasing and field increasing parts of the cycle, i.e.

$$J_c = \frac{3}{2} \frac{M_- - M_+}{R} \qquad (4.4)$$

Here, the factor 3/2 characterises the long cylinder geometry. As shown in chapter 9, the saturation magnetisation takes a value that depends on the geometry of the sample and this coefficient has to be reviewed in each case.

- In view of equation (4.4), when the magnetisation loop displays the behaviour shown to the right of figure 4.3, it means that J_c is noticeably field-dependent in such a range of applied magnetic field. Still, one can use this expression and obtain $J_c(B)$ by replacing the measured experimental values for the upper and lower branches at each applied field H_a. However, as shown in chapter 9, one has to be careful because for finite samples $\mu_0 H_a$ may strongly differ from B,[6] and then, the experimental curve does not correctly represent the real $J_c(B)$ curve.

Some final remarks on inductive measurements

Inductive measurements in bulk superconducting materials may be performed with the above described *extraction method*. As for the case of bulk samples, if the typical signal is big enough, one does not need to resort to ultimate sensitivity techniques. Just for completeness we want to mention that measurements of samples with feeble signals (tiny crystals in fundamental studies, for instance) may require specialised instrumentation as VSM (vibrating sample magnetometers) or even SQUID magnetometry (based on superconducting quantum interference, see section 3.4.2). In case of interest, the reader is directed to [9] that offers a didactic introduction to modern measurement techniques.

Also, we want to mention that figure 4.3 is just an example of what can be done through inductive measurements in superconducting materials. In fact, one could also record M_V as a function of temperature for a given applied field, and determine T_c as the value above which M_V goes to zero. Thus, we have complimentary experiments that may provide the same superconducting parameters, as both T_c and J_c may be either obtained by transport or inductive measurements. Noticeably, the discrepancy of the results obtained either by one method or the other may shed light

---

[5] Recall that the *critical currents* describe the magnetic field gradient caused by non-uniform vortex densities (section 3.3), but they may be accompanied by other contributions, viz.: the equilibrium magnetisation.

[6] Because the contribution of the superconducting sample currents to the total field ($B = B_a + B_{SC}$) may be relevant.

on the microstructural properties of the samples. Thus, a filament with a weak superconducting junction in the middle could become resistive at much lower currents than those evaluated by inductive methods that would detect the super-position of the two halves.

4.3 Magneto-optics

Magneto-optical imaging (MOI) of magnetic materials is based on Faraday's rotation. In the so-called magneto-optical indicators, a rotation of the polarisation of propagating light occurs when a magnetic field is applied [10]. Then, by placing a thin indicator above the surface of a superconductor, and observing the resulting change of the polarisation of light one may obtain information about the underlying magnetic structure. In particular, for superconducting materials under appropriate conditions[7], the experimental data allow to obtain the magnetic field, and by inversion of data, the profile of circulating current densities in the sample. Even more, specialised high resolution techniques have made it possible to visualise individual vortices in type-II superconductors. Here, we will describe the funda-mental principles of the powerful MOI technique. In chapter 11, we provide the tools that allow to obtain the picture of superconducting currents based on the processing of experimental data. For an extensive review concerning the abilities of MOI one may go through the book by Johansen and Shantev [11]. As explained there, after years of refinement, the technique has achieved such advances as resolving the scale from a few cm to sub-microns. Here, our goal is a brief introduction to the general aspects.

4.3.1 Basic experiment: analysis

The basic equipment required for MOI experiments is shown in figure 4.4[8]. A beam produced by a light source is linearly polarised and directed towards a magneto-optical layer (MOL), that is placed above the superconductor. The sample is cooled, typically by using a cold finger in a cryostat. The magnetic field is applied through a coil, typically in the range of tens or at most hundreds of mT. In general, the experiments proceed in reflection mode, which typically requires the use of mirror layers in the lower part of the MOL. Owing to the Faraday effect, the polarisation vector of emerging light is rotated through an angle α, that may be approximated by

$$\alpha \approx V\ell B_z \qquad (4.5)$$

with V being the so-called Verdet constant for the indicator material, ℓ its thickness and B_z the component of the magnetic field along the direction of light propagation.

Apparently, the measurement of α is a direct measure of the intensity of magnetic field in the point of reflection of light. For this purpose, one uses the crossed analyser

---

[7] As emphasised by Brandt in [11], the inversion of magnetic field data does not always have a unique solution.
[8] Extensive details on the actual MOI materials used for investigating superconductors, system specifications, etc may be found in [11] and [12].

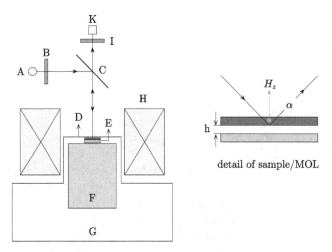

Figure 4.4. Sketch of a conventional setup in magneto-optics. A: light source, B: polariser, C: semi-reflecting mirror, D: optical window of the cryostat, E: sample covered with MOL, F: cold finger, G: cryostat, H: coils, I: analyser, K: CCD camera. To the right a detail of the sample (nonrealistic incidence angle for visual purposes).

and applied Malus' law, that gives the intensity of light, I beyond the analyser in terms of the rotation angle:

$$I = I^{'} \sin^2(\alpha - \Delta\alpha) \tag{4.6}$$

Following the notation in [12], $I^{'}$ represents the intrinsic light intensity emerging from the MOL itself and $\Delta\alpha$ the deviation of the angle between the polarizer/analyser crossing angle from $\pi/2$. Ideally, by combining equations (4.5) and (4.6), one obtains[9]

$$B_z(x, y) = \frac{1}{V\ell}\left[\sin^{-1}\left(\sqrt{\frac{I(x, y)}{I^{'}(x, y)}}\right) + \Delta\alpha\right] \tag{4.7}$$

Thus, by mapping the intensity recorded in the CCD camera at each point of the surface $I(x, y)$, one gets a map of the magnetic field component perpendicular to the surface of the sample.

4.3.2 Quantitative aspects

Two questions arise related to the application of equation (4.7) for the analysis of experimental data

- On the one side, one needs to know the function $I^{'}(x, y)$. This may be achieved by calibration against a given profile of magnetic field, which one knows *a priori*. Notice that this function only depends on the attenuation of the incident beam by the MOL, independent of the rotation angle.

---

[9] Equation (4.7) implies assuming ideal conditions in the experiment. In practice, it has to be modified so as to account for such effects as background signals or MOL anisotropy.

- On the other side, the recorded intensity $I(x, y)$ depends on the absolute value of B_z, or equivalently, the sign of B_z is not determined by equation (4.7). Differential methods, based on the comparison of measurements with deviation angles $\pm\Delta\alpha$ are used to overcome this problem if necessary [12].

4.3.3 Interpretation of results

In its initial stages MOI was basically a qualitative technique that gave information about local aspects of the electromagnetic properties in superconducting materials. One could identify the penetration of magnetic field along preferred directions, the influence of defects, of the actual geometry of the sample, etc. However, quantitative studies have been enabled by two main factors: (i) the improvement of calibration techniques for the light intensity contrast and (ii) the evolution in the understanding of the inverse problem, i.e. for designing configurations that may uniquely determine the internal fields in the sample by measuring close to their surface.

As an example of what may be obtained we include figure 4.5. Actually, the plot is a simulation, but such figures are typically found in experimental works [12]. Thus, the polarised light picture reveals the intensity of the magnetic field. Brighter areas correspond to highest intensity and darker to the lowest (zero in this case). This shows that, when a magnetic field is applied perpendicular to the plane of the platelet, penetration takes place basically along the central part of the sides, defining a so-called cushion-like pattern. On the other hand, as said above, by virtue of the possibility of accurate calibration of the measured intensity, one can *invert* Biot–Savart's law

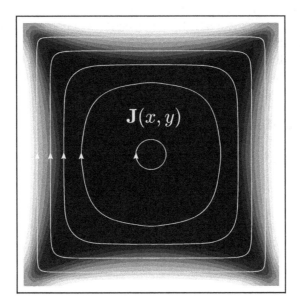

Figure 4.5. Simulation of a typical MOI experiment on a square superconducting platelet. The platelet lies on the XY-plane and a uniform magnetic field is applied along the Z-axis. The intensity of the full magnetic field is revealed from the intensity of light across the analyser.

$$\mathbf{B}(\mathbf{r}) = \frac{\mu_0}{4\pi} \int \frac{\mathbf{J}(\mathbf{r}') \times (\mathbf{r} - \mathbf{r}')}{\|\mathbf{r} - \mathbf{r}'\|^3} \, dV' \tag{4.8}$$

and derive the current density profile $\mathbf{J}(\mathbf{r}')$ on the basis of the $B_z(x, y)$ map above the surface. The method will be explained and illustrated with practical examples in chapter 11. Figure 4.5 shows the corresponding streamlines for the current density in an early stage of field penetration for the square platelet. Having the capacity of performing these kinds of measurements at different values of the applied magnetic field and temperature, MOI constitutes a powerful technique to reveal the fundamental electromagnetic processes in superconducting samples. Noteworthily, MOI also provides a valuable tool, not only in fundamental studies, but also in applied superconductivity. As an example of this, A V Pan and co-workers could confirm that soft ferromagnetic sheaths increase the maximum current capacity in superconducting wires, by concentrating flux in the sheath and producing *overcritical* $(J > J_c(H_a))$ behaviour within the core of the wire (contributed paper in [11]).

4.4 Force measurements

The interaction between superconducting materials and magnets has raised much interest for several reasons. In particular, it is to be mentioned that the interacion between a superconductor and a permanent magnet that approaches the former gives way to levitation forces[10]. Certainly, this is the most popular manifestation of superconductivity since high-T_c materials are available. From the point of view of the technological applications, levitation forces are the focus in many engineering projects focusing on the design of frictionless bearings and vehicles that take advantage of this phenomenon. Thus, the electromechanical characterisation of superconducting samples (arising forces, stiffness coefficients, etc) is a widely used technique. As it will be shown later (chapter 13), although 'practical' materials with acceptable performance are already available, their behaviour under specific conditions is so complex that experimental assessment is crucial. In particular, the actual forces between magnets and superconductors strongly depend on such factors as geometry and the cooling route of the superconductor in the magnetic environment.

From the technical point of view, the characterisation of materials for levitation applications belongs to the macroscale and uses load cells in the range of $F \simeq 10^3$ N. On the other side, force measurements in the microscale have also provided useful information on superconducting parameters for the material. In particular, they have been used to determine the London penetration depth λ_L. In such case, the range of involved forces is $F \lesssim 10^{-10}$ N, a fact that requires specialised experimental techniques and analysis. Below, we will give a basic description of the operation and related physical information for each case.

---

[10] In its basic form this is as follows: induced by the presence of the magnet, supercurrents appear so as to cancel the magnetic field. The associated magnetostatic interaction produces a force that tries to counter balance the presence of the magnet.

4.4.1 Macroscopic force measurements

The typical setup used for the evaluation of levitation properties in practical[11] superconductors is shown in figure 4.6. Basically, one needs a load cell that may be attached to a liquid nitrogen vessel (cryostat) with the superconducting sample firmly clamped inside. Underneath this, the permanent magnet, also firmly secured at the end of a mobile holder is lead along a given path during the experiments. Meanwhile, the load cell records the arising force. In general, 2D measurements are conducted, i.e. $F_x(x,y)$, $F_y(x,y)$ with $[x(t),y(t)]$ being some trajectory of the magnet. Some studies also include fully 3D behaviour. The accurate control of displacement is typically achieved by laser based sensors.

Although not exclusive, a large number of characterisation experiments have been performed by using melt textured superconducting YBCO blocks and NdFeB permanent magnets, all of them with typical dimensions in the range of cm × cm × cm. For separation distances in the range of a few mm, this leads to interaction forces of several hundreds of N.

Directly related to the envisaged application of frictionless levitation vehicles, the components of the force are typically named after levitation (F_z), and guidance (F_x). The reason is apparent, a repulsive F_z is responsible of flotation, and F_x, that should have a restoring character ($\delta x > 0 \Rightarrow F_x < 0$) will ensure stability for lateral disturbances of the vehicle. Just to illustrate the typical output for such measurements, we include figure 4.7, that shows the levitation and guidance forces arising when the magnet is either shifted upwards/downwards from an initial position given

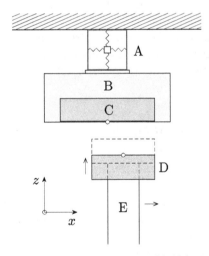

Figure 4.6. Sketch of a magnetic force measurement system with high-T_c superconductors. A: load cell, B: liquid nitrogen cryostat, C: superconductor, D: magnet, E: mobile holder connected to step motors. The lower/upper position of the magnet and the superconductor may be reversed depending on the actual facility.

[11] To date, large scale applications under development are mostly based on high-T_c materials, which may operate at liquid nitrogen temperature (77 K).

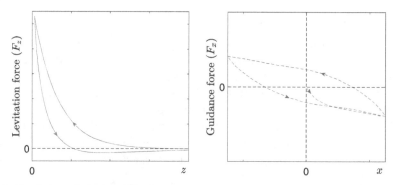

Figure 4.7. Simulation of characteristic force measurements between a superconducting block and a permanent magnet that is moved around for different routes (see text). According to the sketch in figure 4.6, **F** is exerted on the superconductor.

by $(0,a)$, or rightwards/leftwards from an initial position given by $(0,0)$ in the setup shown in figure 4.6. For this figure, the coordinate system settles centred on the top of the magnet and (x,z) indicates the position of the lower point marked on the bottom of the cryostat. Notice that when the magnet comes from a long distance z below the superconductor, the vertical force is repulsive, and increases as distance diminishes (as far as displacement is not reversed!): flotation of the superconductor is induced. On the other hand, as shown to the right of the plot, if the magnet was initially standing just below the superconductor and shifts to the right ($\delta x > 0$), a negative restoring force arises on the superconductor, that would push it towards the original position on top of the magnet (again, while reversal does not occur). It is apparent that for an arbitrary route of the magnet, a very rich phenomenology could be observed, with transitions between stable/unstable points, variations of the magnetic force, etc. These properties strongly change for relative displacements in the region of small distances. Simulation resources for the quantitative analysis are provided in chapter 13.

4.4.2 Magnetic force microscopy: experiment

As said, the evaluation of magnetic forces between magnets and superconductors in the microscopic scale has also raised much interest. As will be explained in chapter 12, based on these measurements, one may obtain the value of a fundamental superconducting parameter, i.e. the penetration depth λ_L. Recalling that, though changing for material to material, λ_L takes values in the scale of tens or hundreds of nm, one may understand that accurate information must be obtained by working with high quality small samples, typically crystals or thin films with thicknesses in the submicron range. The required technology, which takes advantage of the advances in nanolithographic advances is based on the so-called atomic force microscopy (AFM). Basically, as shown in figure 4.7, MFM is based on the detecting the deflection of a tiny cantilever, on the end of which a sharp tip covered with some ferromagnetic material has been deposited. Typical sizes of tens of nm are usual for such tips. The cantilever deflection that is detected by optical methods arises due to the force between the tip and the

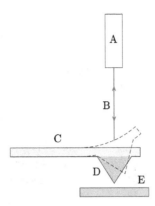

Figure 4.8. Sketch of magnetic force microscopy detection system. A: optical fibre, B: laser beam, C: cantilever, D: magnetic tip, E: superconducting sample

underlying superconductor, typically for separation distances of the order of 1 μm or less. Apparently, in such scales, the recorded forces are sensitive to the typical length along which supercurrents change, i.e. λ_L, as has been demonstrated in recent experiments [13][12]. Nevertheless, the actual dependence may be rather complicated. In chapter 12, we will introduce some tools that may be used for analysing the related inverse problem, i.e. obtain $\lambda_L(F_{MEASURED})$ (figure 4.8).

Other techniques

In addition to a number of modifications to the above mentioned techniques, ranging from the ac extensions of transport and inductive methods, the powerful SQUID magnetometry as an alternative to the latter, hall-probe scanning measurements instead of MOI, or torque magnetometry by similarity to MFM, one has to mention a fundamental family of experiments, i.e. thermal measurements (heat capacity) that have been fundamental in phenomenological and basic superconductivity and occupy a central position in the development of the discipline. Because the scope of this book relates to the electromagnetic area, we have not included them in this chapter.

References

[1] Kamerlingh Onnes H 1911 Further experiments with liquid helium. C. On the change of electric resistance of pure metals at very low temperatures, etc. IV. The resistance of pure mercury at helium temperatures *Comm. Phys. Lab. Univ. Leiden* **120b**

[2] Kamerlingh Onnes H 1911 Further experiments with liquid helium. D. On the change of electric resistance of pure metals at very low temperatures, etc. V. The disappearance of the resistance of mercury *Comm. Phys. Lab. Univ. Leiden* **122b**

---

[12] In fact, the most sensitive technique relies on detecting the variation of the cantilever's resonance frequency by action of the magnetostatic force.

[3] Kamerlingh Onnes H 1911 Further experiments with liquid helium. G. On the electrical resistance of pure metals, etc. VI. On the sudden change in the rate at which the resistance of mercury disappears *Comm. Phys. Lab. Univ. Leiden* **124c**

[4] Murase S, Itoh K, Wada H, Noto K, Kimura Y, Tanaka Y and Osamura K 2001 Critical temperature measurement method of composite superconductors *Physica C: Supercond. Appl.* **357-360** 1197–200

[5] Meissner W and Ochsenfeld R 1933 Ein neuer effekt bei eintritt der supraleitfhigkeit *Naturwissenschaften* **21** 787–8

[6] Hein R A, Francavilla T L and Liebenderg D H 1991 *Magnetic Susceptibility of Superconductors and Other Spin Systems* (New York: Plenum Press)

[7] Brandt E H and Mikitik G P 2002 Why an ac magnetic field shifts the irreversibility line in type-II superconductors *Phys. Rev. Lett.* **89** 027002

[8] Campbell A M and Evetts J 1972 Critical Currents in Superconductors *Monographs on Physics: Taylor and Francis* (London: Taylor & Francis)

[9] Fiorillo F 2010 Measurements of magnetic materials *Metrologia* **47** S114–42

[10] Galperin Y M 2014 *Introduction to Modern Solid State Physics* (Scotts Valley, CA: Create Space Independent Publishing Platform)

[11] Johansen T H and Shantsev D V 2004 *Magneto-Optical Imaging. II. Mathematics, Physics and Chemistry vol 142 Nato Science Series* 1

[12] Jooss C, Albrecht J, Kuhn H, Leonhardt S and Kronmüller H 2002 Magneto-optical studies of current distributions in high-T_c superconductors *Rep. Prog. Phys.* **65** 651–788

[13] Nazaretski E, Thibodaux J P, Vekhter I, Civale L, Thompson J D and Movshovich R 2009 Direct measurements of the penetration depth in a superconducting film using magnetic force microscopy *Appl. Phys. Lett.* **95** 262502

Part II

Mathematical tools and computation

Chapter 5

Some useful mathematical resources

This chapter is aimed at reviewing some mathematical tools that will be extensively used in the quantitative study of superconducting phenomena. Although the reader will be provided with numerical tools for each of the case studies, a coherent connection between the physical concepts and the calculations requires a good establishment of the mathematical techniques to be used. Also, this is a key feature for further development.

A basic knowledge of differential calculus, vector operations and matrix algebra is assumed and is not focused on here. Specifically, we will concentrate on two methods that may be less familiar: the calculus of variations and the use of discrete formulations. A rather elementary approach (what is needed later, basically) will be made, mostly based on dedicated examples. Those who are already familiar with these topics might skip this chapter. In any case, they are invited to go through so as to get familiar with the notation, that will be used in the rest of the book.

5.1 Variational calculus

The calculus of variations is a powerful branch of mathematics with useful applications in several areas of physics, engineering and economy. As concerns physical modelling, it is the background technique to solve minimum principle statements, omnipresent in nearly every branch of the discipline. Just to quote a couple, we mention the minimisation of energy, and the least action principle, that will be repeatedly applied in this book. In this section, we offer a very simplified introduction, aiming at a first contact for those who have no practice with this method, and also a reminder for those who already have. Our exposition, as concerns the theoretical background, is partly inspired by contents in the books by Arfken [1] and Elsgolts [2], that may serve as further reference for the interested reader. The numerical implementation for solving a collection of problems in applied super-conductivity will be done by relying on MATLAB and its own optimisation tools (see section 6.1.6). For those who have want to tackle large scale problems, a recommended reference is the FORTRAN code package GALAHAD [3].

5.1.1 The basic problem: concept of variation

In its basic form, variational calculus states the following problem: find the minimum[1] of the integral

$$S[f(x)] = \int_{x_A}^{x_B} L\left(f, \frac{df}{dx}, x\right) dx \tag{5.1}$$

with f being an unknown function of the variable x that satisfies the boundary conditions:

$$\begin{cases} f(x_A) = f_A \\ f(x_B) = f_B \end{cases} \tag{5.2}$$

S, i.e. the quantity that must take a stationary value, is a so-called *functional*, and the integrand L is a certain combination of f, its derivative f' and the independent variable x.

5.1.1.1 Example 1

As a popular introductory example consider the *geodesic* problem in the Euclidean plane, i.e. find the path that joins points A and B which has the shortest length. Certainly, it is a straight line. Let us see how this arises in the formulation. As illustrated in figure 5.1, among an infinite number of curves $y(x)$ that join such points, one must choose the one along which

$$S[y(x)] = \int_{x_A}^{x_B} \sqrt{dx^2 + dy^2} = \int_{x_A}^{x_B} \sqrt{1 + \left(\frac{dy}{dx}\right)^2} \, dx \tag{5.3}$$

is a minimum. Notice that, in this case, the function $f(x)$ takes the value of the y coordinate, so that $dx^2 + dy^2$ is the squared length of an infinitesimal displacement.

Conceptually, one searches the function $y^*(x)$ that produces a minimum in equation (5.3). A brute force method to find it would consist of choosing a certain trajectory $y^{(n)}(x)$, comparing to another, disregarding that one that produces the bigger value and so on, expecting to achieve some convergence. Systematically, the problem is solved as explained below.

One proceeds in a parallel fashion to the standard search of a criterion that characterises the appearance of minima for a function in differential calculus. We define the 'variation' $\delta f(x)$ that measures the difference between a trial function and some reference function that satisfies the boundary conditions. Assuming differentiability, the necessary condition for $f^*(x)$ to produce an extremum in

---

[1] In general, one searches for stationary points, i.e. minima, maxima or inflection points. In physics one is very frequently focused on minima.

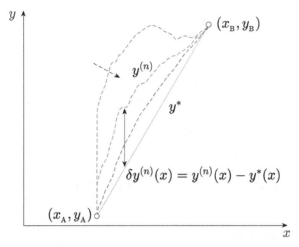

Figure 5.1. Illustration of the variational *geodesic* problem. The minimum length path between points *A* and *B* is searched. Several trajectories differing less and less from the solution are displayed.

$S[f(x)]$ is that arbitrary variations produced by 'testing' around such a function go to zero: $S[f^*(x) + \delta f(x)] - S[f^*(x)] \to 0$. It is shown that this leads to the following condition to be satisfied by the function candidate to give a minimum:

$$\frac{\partial L}{\partial f} = \frac{d}{dx}\left(\frac{\partial L}{\partial f'}\right) \tag{5.4}$$

This is the so-called *Euler–Lagrange* equation. Let us apply it to our example: $L = \sqrt{1 + y'^2}$. Equation (5.4) leads to the condition

$$\frac{y''}{\sqrt{1 + y'^2}} = 0 \quad \Rightarrow \quad y' = \text{constant} \tag{5.5}$$

which apparently gives the expected result. The actual expression $y(x)$ may be determined by imposing the boundary conditions (5.2).

We notice, in passing, that for our purposes the Euler–Lagrange equations will be basically a method for reassuring the theoretical background of the physical models. Thus, in chapter 1 (section 1.2.4) we have used them to show that one may issue a variational form of Faraday's law for conducting materials. On the other side, the practical application of variational statements will be performed by numerical means. As an example of performance, consider that the above *minimum distance* problem may be solved as follows. Discretise the region of interest by a set of points $\{x_i\}$ with $i = 1, \ldots, N$ and $x_1 = x_A$, $x_N = x_B$, and search for the set $\{y_i\}$ satisfying $y_1 = y_A$, $y_N = y_B$ while the discrete version of the functional, i.e.

$$S = \sum_{i=1}^{N-1} \sqrt{(x_{i+1} - x_i)^2 + (y_{i+1} - y_i)^2} \tag{5.6}$$

is minimised.

In fact, figure 5.1 has been obtained in this fashion with help of the MATLAB function fmincon() (see section 6.1.6 for details on this function). The different curves correspond to the application of the numerical algorithm for increasing number of iterations, until convergence (the solution no longer changes by increasing the number of iterations) to the straight line is observed.

As a final remark of this section, we comment that the statement of minimising (5.1) and the corresponding Euler–Lagrange equations (5.4) may be generalised in several respects. In particular, the theory may be extended to functionals depending on several functions of several independent variables, as the reader may find in basic references such as [1] and [2]. Roughly speaking, the above considerations are valid and one obtains equations in more variables as will be shown in the following example.

5.1.2 Constrained variations

A different generalisation of the variational problem that will be essential in our physical applications is the consideration of constraints on the searched solution (additional to the boundary conditions). Several types of constraints may be considered.

5.1.2.1 Example 2
One may need to minimise the physical functional with a constraint of the type $\varphi(x, f, f') = 0$, i.e. some implicit restriction on the searched function f. Let us see how this applies with an example of magnetostatics, the minimisation of magneto-static energy. The role of f will be played by the vector function $\mathbf{B}(\mathbf{r})$. The constraint will be that Ampère's law must be satisfied, i.e.

$$\varphi = \frac{\nabla \times \mathbf{B}}{\mu_0} - \mathbf{J} = 0 \tag{5.7}$$

with $\mathbf{J}(\mathbf{r})$ prescribed by the sources of the problem (a set of coils for instance).

According to theory (5.7) may be imposed by the so-called Lagrange multiplier method, which implies to use an *augmented* functional that includes the constraint. Thus, the energy term $B^2/2\mu_0$ will be supplemented by a term of the form $\lambda \cdot \varphi$ with λ being the above mentioned Lagrange multiplier. We end up with the functional

$$S[\mathbf{B}(\mathbf{r}), \lambda(\mathbf{r})] = \int \left[\frac{B^2}{2\mu_0} + \lambda \cdot \left(\frac{\nabla \times \mathbf{B}}{\mu_0} - \mathbf{J} \right) \right] dV \equiv \int \mathcal{L} \, dV \tag{5.8}$$

Now, both $\mathbf{B}(\mathbf{r})$ and $\lambda(\mathbf{r})$ are dependent variables to be determined by Euler–Lagrange equations. For the former, they read[2]

$$\frac{\partial \mathcal{L}}{\partial \mathbf{B}} = \frac{\partial}{\partial x} \frac{\partial \mathcal{L}}{\partial (\partial \mathbf{B}/\partial x)} + \frac{\partial}{\partial y} \frac{\partial \mathcal{L}}{\partial (\partial \mathbf{B}/\partial y)} + \frac{\partial}{\partial z} \frac{\partial \mathcal{L}}{\partial (\partial \mathbf{B}/\partial z)} \tag{5.9}$$

---

[2] Notation: bold quantities are used to indicate that we have as many equations as components of $\mathbf{B}$.

which one may check that leads to

$$\mathbf{B} = \nabla \times \lambda \qquad (5.10)$$

i.e. the Lagrange multiplier λ may be identified with the vector potential ($\mathbf{B} = \nabla \times \mathbf{A}$). On the other hand, considering λ as a second dependent variable in the functional, one may straightforwardly check that one is lead to Ampère's law by imposing the corresponding Euler–Lagrange equation.

Concerning the implications of the above analysis for numerical purposes, it can be said that minimising the augmented functional (5.9) one implicitly includes Ampère's law in the formulation.

5.1.2.2 Example 3

A second kind of constraint that will be significant for the kind of problems analysed in this book are the so-called *one-sided* or *unilateral* constraints. Figure 5.2 shows a simple situation that illustrates such a concept. Consider again the *geodesic* problem of the first example, but assume that a forbidden region exists. For instance, imagine that a circular hole in the ground does not allow to use the straight line connecting points *A* and *C*. What is then the optimal path joining such points? A number of theoretical results [2] could be used to find the solution for this problem, just by using pencil and paper, and simple reasoning. The solution should be composed of straight lines in the allowed region and arcs along the boundary of the hole, connecting smoothly with the former. This is what one can see in figure 5.2, which in fact, was obtained by using MATLAB optimisation function fmincon(). As said (expression (6.3)) the numerical algorithm allows to include restrictions of the kind $C(X) \leqslant 0$, which in this case may be used in the form

$$R^2 - x^2 - y^2 \leqslant 0, \qquad (5.11)$$

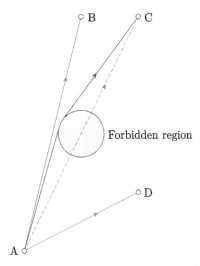

Figure 5.2. *Geodesic* trajectories in Euclidean space, joining the starting point A with endpoints B, C, D. A forbidden region determines that the trajectory A → C differs from the straight line.

to imply that our solution 'rejects' the points within the circle of radius R. Recall that fmincon() gives the correct solution for the situations $A \to B$, $A \to C$, $A \to D$, i.e. the algorithm determines whether constraints must be active or not. This kind of problem straightforwardly connects to the so-called critical state problem in type-II superconductivity (section 3.3). In such case, we will search for the current density distribution in a superconducting material for which a restriction of the kind

$$J_x^2 + J_y^2 + J_z^2 \leqslant J_c^2 \tag{5.12}$$

will be imposed, as related to the underlying physical mechanisms.

5.1.3 Non-smooth solutions

A simple example, illustrated in figure 5.3, shows that, in some instances, minimisation is realised by non-smooth functions. In our case, searching for the behaviour of physical quantities, one must be ready to argue about the validity and interpretation of the related mathematical model. Let us use this example for clarification. In figure 5.3 we display the result of minimising the functional

$$S[f(x)] = \int_{x_A}^{x_B} f^2 \, dx \tag{5.13}$$

for the boundary conditions $f(x_A) = f(x_B) = f_0$.

Apparently, the 'function' given by

$$f^*(x) = \begin{cases} f_0, & x = x_A \\ 0, & x_A < x < x_B \\ f_0, & x = x_B \end{cases} \tag{5.14}$$

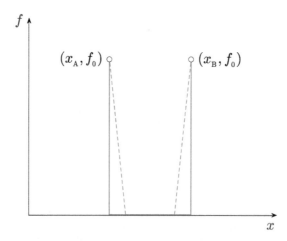

Figure 5.3. In some statements, minimisation is realised by discontinuous functions. Continuity may be recovered by imposing restrictions as in equation (5.15).

minimises (5.13), giving $S[f^*(x)] = 0$. Mathematically, f^* is discontinuous, but obviously it is the solution of the problem. Physically, one should be cautious to accept this. Could such a function represent the magnetic field being expelled from a superconductor? In order to reconcile both aspects, let us imagine that the function f is subject to some restriction. What if one requires that the derivative of the function is bounded?

$$\left|\frac{df}{dx}\right| \leqslant g_0 \qquad (5.15)$$

Then, minimisation is realised by a continuous function as indicated with dashed lines in the plot, i.e. when the derivative takes extremal values allowed $f' = 0, \pm g_0$.[3] Let us see how the statement in (5.13) may acquire a physical interpretation. If one replaces f by the magnetic field $\mathbf{B}$, f_0 may be interpreted as the prescribed value B_a of an applied magnetic field that is expelled from a superconducting slab of thickness $x_B - x_A$ and the statement correspond to the minimisation of magnetostatic energy. The non-smooth solution must be understood as a limiting case that corresponds to a high value of the maximum current density allowed in the material ($g_0 \to \infty$ in (5.15)). If g is bounded, one gets the profile shown in the plot.

5.2 Discrete formulation

The numerical solution of electromagnetic problems is a well developed discipline with numerous applications, amongst other superconducting elements. Highly sophisticated methods have been issued, based on solid theoretical background [5], and already at the level of practical implementations in highly complex systems [6]. Here, aiming at a first contact with the problem, we will introduce some elementary ideas. Such concepts as finite or vector element methods would be a natural continuing step. Thus, for our purposes, a so-called finite difference approach will be used. The quantities of interest will be evaluated on the points of a regular mesh, and their derivatives (if necessary) approximated by differences involving the neighbouring points. As remarked by Bossavit [5], finite difference schemes perform rather well for regular regions, which will be our focus. Those who have interest in regions with complicated geometries should consider the use of specialised finite elements[4].

5.2.1 Grids and regions

The first task in the formulation of a discretised problem will be to define a grid of points as support for evaluating the physical quantities of interest. Although working with vector fields (recall that our objects of interest are $\mathbf{B}(\mathbf{r})$, $\mathbf{J}(\mathbf{r})$, $\mathbf{E}(\mathbf{r})$) all the problems treated in this book will allow some simplification that permits to

---

[3] This result is a particular case of the very general result called *the bang-bang principle* of the *optimal control* theory [4], a powerful generalisation of the calculus of variations.

[4] Along this book we will refer to finite elements, though as said the concept is much wider and we will restrict to the simplest finite difference approximation.

state them in terms of some scalar field evaluated in the mesh. As said, we will focus on regular domains such as those shown in figures 5.4 and 5.5. In order to illustrate the actual implementation, we will introduce an example based on a physical system that will be studied in chapter 10.

5.2.1.1 Example

Consider a (super)conducting square platelet subjected to a perpendicular magnetic field. The platelet occupies the region Ω given by the conditions $|x| \leqslant a$, $|y| \leqslant a$, $|z| \leqslant w \ll a$. The magnetic response of the system will be characterised by a surface current density $\mathbf{K}(\mathbf{r})$ that one may express in terms of a scalar function[5] $\sigma(x, y)$

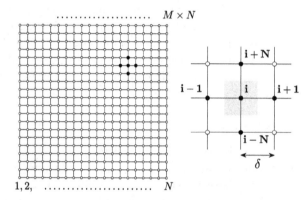

Figure 5.4. Definition of a rectangular grid. To the right, we highlight a generic node of the grid, i, its nearest neighbours, and an elementary area around it.

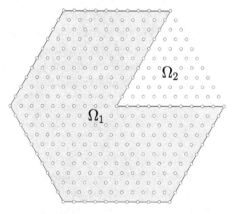

Figure 5.5. Discretisation by a regular triangular mesh. The full problem is composed of two subregions (Ω_1, Ω_2) with regular shapes.

---

[5] As explained in chapter 10.

$$\mathbf{K} = -\hat{z} \times \nabla\sigma \tag{5.16}$$

Among other properties, we will evaluate the sample's magnetic moment in terms of σ:

$$m_z = \int_\Omega \sigma(x, y)\, dxdy \tag{5.17}$$

The function $\sigma(x, y)$ will be approximated by its values at a rectangular grid as that one shown in figure 5.4 (in this case $M = N$). Recall that the grid defines an array of nodes, each one labelled with an index i to a total amount of $n = N \times M$. Apparently, the bigger the value of n, the smoother the representation, but also the computation times.

Thus, we will work with the collection of values of $\sigma(x, y)$ at the nodes, i.e. $\sigma_i = \sigma(x_i, y_i)$. As physical dimensions must be considered for quantitative purposes, an important parameter of the model will be the distance δ defined in the sketch, i.e. the distance between two nodes

$$\delta = \frac{a}{N} \tag{5.18}$$

This gives a conversion factor between actual calculations and the specific sample under consideration. Customarily, we will use dimensionless units

$$\begin{aligned} \tilde{x}_i &\equiv \frac{x_i}{\delta} = 1, \dots, n \\ \tilde{y}_i &\equiv \frac{y_i}{\delta} = 1, \dots, n \end{aligned} \tag{5.19}$$

and eventually apply the dimensions of the system by multiplying by δ.

Concerning the evaluation of quantities derived from the set of values $\{\sigma_i\}$, as those mentioned above, one proceeds as follows.

Derivatives. First order derivatives may be approximated by differences. As sketched in 5.4, one may use[6]

$$\begin{aligned} \frac{\partial\sigma}{\partial x} &\approx \frac{\sigma_{i+1} - \sigma_i}{\delta} \quad \text{at} \quad (x_i, y_i) \\ \frac{\partial\sigma}{\partial y} &\approx \frac{\sigma_{i+n} - \sigma_i}{\delta} \quad \text{at} \quad (x_i, y_i) \end{aligned} \tag{5.20}$$

Integration. The evaluation of expression (5.17) may be done as

$$m_z \approx \sum_{i=1}^n \sigma_i A_i = \sum_{i=1}^n \sigma_i \delta^2 \tag{5.21}$$

Notice that the elementary area corresponding to the nodal point i is a square of area δ^2 (as emphasised in the plot).

---

[6] For points at the boundary of the region, one must take care so as to avoid going beyond the limits.

Some eventual comments are due. On the one hand, the generalisation of the above expressions to a rectangular surface ($a \times b$) is obvious, just by introducing $\delta_n = N/a$, $\delta_m = M/b$. On the other hand, the extension to non-rectangular grids is more intricate, both for the evaluation of derivatives and integral expressions. For instance, one may need to use a node-dependent expression for the element of area A_i as discussed in chapter 9 (Figure 9.5).

As an example of another situation of interest that also allows a regular mesh, we include figure 5.5, that displays a triangular grid fitting regular regions composed of equilateral triangles. Notice in passing that a given discretisation may be used to cover inhomogeneous behaviour if some degree of symmetry is maintained.

If the region of interest may be described as the superposition of different parts, we will use the mathematical concept of *direct summation*. The representation of the material will be done by the 'juxtaposition' of vector components representing the state of each of the constituent parts. As indicated in figure 5.5, if the section of the sample consists of two parts, say $\Omega = \Omega_1 \bigcup \Omega_2$, one will solve for the physical quantity in each of them and characterise the overall state by the set of values $\{\sigma_I\}$ with $I = 1, \ldots, n_1; n_1 + 1, \ldots, n_1 + n_2$, and n_1, n_2 the full number of grid points in the respective subregions. We will use the notation: $\sigma_{\text{FULL}} = \sigma_1 \oplus \sigma_2$.

This concept will be helpful when dealing with mixed material structures for which the subdivision is apparent, or even for homogeneous materials in which symmetry properties allow to generate a solution for the problem by replicating some subset of values (see chapter 14).

5.2.2 Vectors and matrices

A rather convenient representation of the physical quantities and operations with them on a grid consists of 'gathering' the values of the former in vector form, and using matrices for the linear operators. Continuing with the above example, and making use of Dirac's 'ket' and 'bra' notation, we define the vectors and their transposes

$$|\sigma\rangle \equiv \begin{pmatrix} \sigma_1 \\ \sigma_2 \\ \cdot \\ \cdot \\ \sigma_n \end{pmatrix}, \qquad \langle\sigma| \equiv (\sigma_1, \sigma_2, \ldots, \sigma_n) \qquad (5.22)$$

with n being the number of points in the grid ($n = N \times M$). Mathematically, the components of the n-dimensional vector $|\sigma\rangle$ are nothing but the values of the function $\sigma(x, y)$ at the points of the grid: σ_i, $i = 1, \ldots, n$.

It is very important to clearly distinguish between these 'n-vectors' and the physical vector fields in real space $\mathbb{R}^3$, which are indeed our ultimate goal. With this in mind, below we establish the notation for the case of the magnetic field, that may be easily extrapolated to the current density and electric field.

- On the one hand, we use the vector function $\mathbf{B}(\mathbf{r})$ to define the components of the magnetic field in real space $\mathbb{R}^3$

$$\mathbf{B}(\mathbf{r}) \equiv \begin{pmatrix} B_x(\mathbf{r}) \\ B_y(\mathbf{r}) \\ B_z(\mathbf{r}) \end{pmatrix} \tag{5.23}$$

- On the other hand, we use the following notation for the values of the components of the magnetic fields at the points of the mesh. Each component is specified by an n-dimensional vector.

$$|\mathbf{B}_x\rangle \equiv \begin{pmatrix} B_x(\mathbf{r}_1) \\ B_x(\mathbf{r}_2) \\ . \\ . \\ . \\ B_x(\mathbf{r}_n) \end{pmatrix}, \ |\mathbf{B}_y\rangle \equiv \begin{pmatrix} B_y(\mathbf{r}_1) \\ B_y(\mathbf{r}_2) \\ . \\ . \\ . \\ B_y(\mathbf{r}_n) \end{pmatrix}, \ |\mathbf{B}_z\rangle \equiv \begin{pmatrix} B_z(\mathbf{r}_1) \\ B_z(\mathbf{r}_2) \\ . \\ . \\ . \\ B_z(\mathbf{r}_n) \end{pmatrix} \tag{5.24}$$

Recall that Dirac's notation gives rather intuitive expressions for many quantities of interest with 'grid quantities'. For instance, recall that

$$\frac{1}{V} \int_\Omega \mathbf{B}^2 dV \approx \langle \mathbf{B}_x|\mathbf{B}_x\rangle + \langle \mathbf{B}_y|\mathbf{B}_y\rangle + \langle \mathbf{B}_z|\mathbf{B}_z\rangle \tag{5.25}$$

which shows that the magnetostatic energy of the system may be approximated in terms of 'scalar products' of our n-dimensional vectors.

The notation for matrix operations with the n-dimensional vectors will be as in the following example, that indicates that the current density may be obtained by linear operations with elements of the vector $|\sigma\rangle$ (recall expression (5.20))[7]

$$|\mathbf{K}_x\rangle = \mathbf{D}_x|\sigma\rangle; \quad |\mathbf{K}_y\rangle = \mathbf{D}_y|\sigma\rangle \tag{5.26}$$

In conclusion, we call the readers' attention to that many of the calculations in this book will consist of matrix algebra operations: matrix addition, product, inversion, etc and their action on vectors, that are conveniently expressed in Dirac's notation. As an example we mention that the Meissner state of finite size samples will be resolved by inverting a certain geometrical matrix and applying the result on a vector. Thus, in chapter 9, we will show that the distribution of Meissner currents on the surface of a finite sample is conveniently obtained as by:

$$|\mathbf{K}\rangle = -\mathbf{m}^{-1}|\psi_a\rangle \tag{5.27}$$

with $\mathbf{m}$ being the mentioned geometrical matrix and $|\psi_a\rangle$ an n- vector determined by the applied magnetic field at the points of the grid.

Certainly, the size of our matrices will be large, as n may reach hundreds or even thousands in a typical grid, but they can be very efficiently handled with the help of

---

[7] $\mathbf{D}_x, \mathbf{D}_y$ will be $n \times n$ sparse matrices with nonzero values ± 1 around the diagonal. They are explicitly evaluated in chapter 10.

such programming languages as MATLAB or OCTAVE. A dedicated introduction to their usage is included in chapter 6.

References

[1] Arfken G B and Weber H J 1995 *Mathematical Methods for Physicists* 4th edn (New York: Academic)

[2] Elsgolts L 1977 *Differential equations and the Calculus of Variations* (Moscow: Mir)

[3] Conn A R, Gould N I M and Philippe L T 1992 *LANCELOT: a Fortran package for large-scale nonlinear optimization (Release A)* (Heidelberg: Springer)

[4] Knowles G 1981 *An Introduction to Applied Optimal Control* (New York: Academic)

[5] Bossavit A 1998 *Computational Electromagnetism: Variational Formulations, Complementarity, Edge Elements.* Academic Press Series in Electromagnetism (New York: Academic)

[6] Pardo E and Kapolka M 2016 3D computation of non-linear eddy currents: variational method and superconducting cubic bulk *J. Comput. Phys.* **344** 339–63

Chapter 6

Introduction to computational methods

This chapter aims at a basic introduction focused on the usage of the provided computational resources for performing calculations and simulations in macroscopic superconductivity. Experienced MATLAB and OCTAVE users may straightforwardly skip it. The fully documented open source programs attached to part III of this book may be used and personalised with a basic knowledge of these languages.

As a large number of books and web resources exist on this topic, here we will just pick up a number of specific concepts and put them together for the readers' benefit. Concerning additional assistance, one may visit the home pages of both languages. Specific pedagogical resources may be found in https://matlabacademy.mathworks.com and https://wiki.octave.org.

On the other hand, for those who are having a first contact, we suggest considering the online versions, that allow you to get familiar with many features of their performance and do not require any installation, as they run in most browsers[1]. Practice with the examples provided in this chapter is highly recommended.

6.1 MATLAB: some basics

MATLAB is a world renowned technical software very popular among scientists, academics and engineers. It has been under continuous development since the late 1970s, and although originally conceived as a friendly interface for performing matrix operations, it has become a reference powerful programming language with extensive numerical and also symbolic capabilities. It can operate in nearly every platform. It runs stand-alone in high performance computing clusters and desktop computers, as well as deployed online from a web browser. Some of its hallmarks are:

- A very steep learning curve for the new users owing to a flexible and intuitive syntax.

---

[1] https://es.mathworks.com/products/matlab-online, https://octave-online.net.

doi:10.1088/978-0-7503-2711-4ch6

- The existence of a vast community of users that share their experience through different sites[2].
- A good connectivity with other programming languages[3] and the machine's operating system.

Based on these features, a good number of researchers develop their codes in MATLAB. The numerical resources provided in this book are a collection of programs with the focus on allowing straightforward hands-on experience in a number of problems of superconductivity. In this section, we offer a basic introduction to those features of the language that may be of help for newcomers for our specific purposes. Also it may be considered as a reminder of some resources for those who already have some knowledge. In fact, only those concepts that are of direct application will be reviewed.

This section is basically organised according to a classical tutorial structure. The reader is suggested to straightforwardly perform a series of elementary exercises so as to get familiar step-by-step with the language. The exposition will be conducted by examples of further application in the physical problems to come in part III.

6.1.1 User's interface

In its current versions MATLAB launches a system of windows, the so-called *desktop environment*, that allows to execute simple commands, their combinations (scripts), open and modify files and visualise results[4]. These windows include the so-called *command window, editor, current folder, workspace* and *history*, whose appearance may be personalised by the user.

To start with, when MATLAB opens, within the window named *current folder* one gets a list of preloaded files in a default directory. Such directory may be personalised by the user through the panel of preferences. We recommend to start by creating a directory (folder) in your disk with the files provided for this chapter (see appendix A: Source Codes). Several extensions may be visible:

.m for matlab code files
.fig for matlab figures
.dat for plain data files
.pdf for .pdf files
.mat for files that store workspace information from a previous session

One may straightforwardly right-click on these files for a number of operations, or equivalently type a command in the prompt (>> symbol) of the *command window*.

---

[2] It is recommended to visit https://www.mathworks.com/matlabcentral/fileexchange.

[3] For example, integration with C++, PYTHON and COMSOL is well developed and nearly full portability applies with GNU OCTAVE.

[4] Occasionally, one could prefer to launch Matlab in terminal mode. In Linux systems this may be done from the system prompt: $ matlab -nodisplay.

There are a number of operations that one may perform with files within the mentioned folder. Start by typing:

```
>> open arrow.m
```

This command opens the program file named *arrow.m* (see below) and shows it in the *editor* window. One may notice that syntax is highlighted in the editor: comment lines are green, language commands blue, functions black. In this window one can edit the file, compare it with others, as well as execute the code within, even with help of a very useful debugger.

```
%%%%%%%%%%%%%%%%%%%%%%%%%%%%%%%%%%%%%%%%%%%%%%%%%%%%%%%%%%%%%%%%%%%%%%%%
%                         File:   arrow.m
%
% This function defines an arrow head centered at the point  (y,z)
% wihtin the plane x= 0. It is shaped as a triangle
% The arrow axis is tilted by an angle theta
% Its 'orientation'(inwards/outwards) is given by o =  +1 or -1
% 'scale' s controls its size
% The output (arrow.Y,arrow.Z) are the vertex coordinates
% The output arrow.C is void and will be used to define color
%%%%%%%%%%%%%%%%%%%%%%%%%%%%%%%%%%%%%%%%%%%%%%%%%%%%%%%%%%%%%%%%%%%%%%%%

function [arrow]=arrow(y,z,theta,s,o)

arrow.Y=y+[0.14*s*o*(-cos(theta)+sin(theta))...% line was too long
    0.14*s*o*(cos(theta)+sin(theta))  -0.2*s*o*sin(theta)];

arrow.Z=[(z+0.14*s*o*cos(theta)+0.14*s*o*sin(theta))...
  (z-0.14*s*o*sin(theta)+0.14*s*o*cos(theta))  ...
    (z-0.2*s*o*cos(theta))];

arrow.C=[];              % Initialize field for the color of the arrow
```

Later on, the function `arrow()` will be used to plot arrow heads. (Source codes available at https://iopscience.iop.org/book/978-0-7503-2711-4.)

Another frequent operation, related to file management, is to open or close data files to be used in some operation. Thus, by typing

```
>>C=load('contour.dat');
```

one may see a new entry appearing in the *workspace window*. The variable C, ready to be used for any operation, stores the coordinates of the points that define a certain line of magnetic field as obtained in some calculation. They are stored in a two-column format. Just by dropping the semicolon in the command line one may visualise the actual list of values. One may load it and make a plot, for instance. Conversely, one may decide that the result of a calculation (variable D for instance) should be stored for further use. Use the command >> `save` for this purpose. In fact, in our scripts one frequently finds a command line of the kind

```
>>save('filename.dat','D','-ascii')
```

This creates a file which stores the variable D in -ascii format for portability.

As a final example of essential operations with files through the command line in the command window, we propose to check:

```
>> open figure_6_2.fig
```

This loads a previously created figure in a special figure window. Then, a large number of operations may be performed on the plot as one may discover through the intuitive upper menu. The *.fig* extension indicates that this is a source file, that may be modified at any time and also exported to a number of formats for edition in some document.

It must be stressed that, in addition to file management, the *command window* allows to perform other useful operations. One may enter operative system type commands as follows: (i) >> (un)zip, that decompress, compress files (ii) >> pwd, that shows us the working directory, (iii) >> ls, that displays its contents, or (iv) >>cd ../, that navigates backwards in the directory tree. Also, one may launch the scripts or functions[5] stored in .m files. For instance, if on enters

```
>> v=arrow(0,0,pi/2,1,1);
```

One gets the vertices of a triangle centred at the point (0, 0), and rotated by $\pi/2$. The coordinates of the vertices may be displayed by typing the name of the new variable[6]

```
>> v
v
struct with fields:
Y: [0.1400 0.1400-0.2000]
Z: [0.1400-0.1400-1.2246e-17]
```

Thus, as a result of some calculation (see above) 'v' stores two arrays, one for the first coordinates of the three vertices v.Y and one for the second coordinates v.Z. Their respective values may be obtained by typing v.Y(1) and so on.

Finally, one may also straightforwardly perform operations in the *command window*, by using it in a 'calculator mode'. For instance, one could calculate the angle in the head of our arrow (see figure 6.1). Start by defining the vector components

```
>> y1=(v.Y(1)-v.Y(3));z1=(v.Z(1)-v.Z(3));
>> y2=(v.Y(2)-v.Y(3));z1=(v.Z(2)-v.Z(3));
```

---

[5] Scripts and functions are command sequences that execute one after the other. However, functions are more flexible because they work on arbitrary input values for the variables, according to the basic structure function OUTPUT = f(INPUT).

[6] Alternatively by omitting the semicolon in the previous instruction.

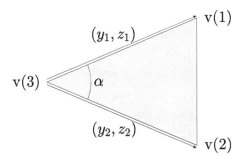

Figure 6.1. A triangular area defined by its three vertices v(1), v(2), v(3). Also marked are the vectors along two edges, with respective components (y_1, z_1) and (y_2, z_2) and the angle in between α.

Then, scalar multiply the vectors and obtain the angle (be aware that by default the output is given in radians).

```
>> s=y1*y2+z1*z2;
>> acos(s/sqrt(y1^2+z1^2)/sqrt(y2^2+z2^2))
ans=
0.7812
```

Here, we have decided just to display the result. One could load the result to some variable just by using the assignment: `>> alpha=acos(..);`

6.1.2 Features of the language

In the previous section, we only offered some notions about how MATLAB® works. Here, we will highlight the features of the programming language that have a special significance in our applications. First, we suggest some practice to get familiar with the essential ideas. Then, some pieces of software, related with problems of basic electromagnetic theory, will help us to illustrate the application.

Data types: numeric variables
As said, MATLAB's syntax is very flexible. Just by using the assignment operator = one gets implicit declaration of the type of variable. What is more, the type is just decided as a result of the calculation. Just to check this idea, one may use the following commands

```
>> z=sqrt(-1)
z =
0.0000 + 1.0000i
>> w=sqrt(z)
w=
0.7071 + 0.7071i
```

The program has implicitly declared that z is a complex number and performed complex arithmetics.

Data types: strings
In some instances, we will make use of string variables. In particular, they will be useful for creating filenames as in the following example

```
>>filename=strcat('file',num2str(#),'.dat')
```

The filename will be the string formed by concatenation of the three strings. The second one is used to label with a numeric value. For instance, if one uses $\# \to 4$ the filename would be *file4.dat*.

Data types: arrays and matrices
Undoubtedly, since its creation, one of the major capabilities of MATLAB is the flexibility and power to work with matrices. Taking advantage of this is something that the user will appreciate. Thus, one may straightforwardly assign array or matrix types to variables and perform operations with them in a very natural and effective manner. As a primer of this, type, for instance[7]

```
>> a=[0 pi/4 pi/2 3*pi/4 pi];
>> b=cos(a)
b =
1.0000 0.7071 0.0000 -0.7071 -1.0000
```

Here, *a* is recognised as a row array and the function cos() is performed automatically element by element. This property is extensive to the long list of pre-defined mathematical functions.

A step further, that introduces some more syntax is to define a matrix. Consider this example

```
>> theta = pi/4;
>> R=[cos(theta) sin(theta); -sin(theta) cos(theta)]
R =
0.7071 0.7071
-0.7071 0.7071
```

This operation creates a 2×2 rotation matrix with angle $\pi/4$. Notice that spaces separate elements in different columns, while semicolons separate rows.

Data types: structures and cells
In some cases one may be interested in using data structures of a more general nature than matrices. Structures and cell arrays allow to deal with inhomogeneous collection of data.

Within the so-called *structures*, one groups related data organised by fields, which may contain different kinds of elements. Stepping back to the script *arrow.m* that generates the data for figure 6.1, one defines a structure with three fields: arrow.Y that contains the sequence with *y*-coordinates of the vertices, arrow.Z that

---

[7]Equivalently, one may use >> a=0:pi/4:pi, whose syntax is apparent.

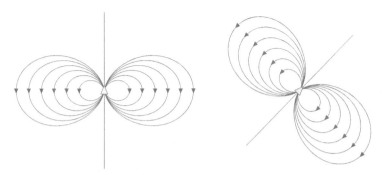

Figure 6.2. Plot of the magnetic field lines around the magnetic dipole for two different orientations 🔗. Source codes available at https://iopscience.iop.org/book/978-0-7503-2711-4.

contains the z-coordinates, and `arrow.C` a container for the colour to be used when plotting the arrow head. Apparently, the nature of these fields may be fully different if desired. For the case of vector quantities, one may consider this as a coherent notation that relates to their components.

In a similar fashion, *cell arrays* are indexed containers that may hold different kind of data in each cell. As a typical example of their usage, we address the reader to the software provided for figure 6.2. We notice that each of the calculated magnetic field lines is nothing but a collection of coordinates (y_i, z_i) with a number of points $i = 1, 2, \ldots, n_J$ that will be possibly different for each line. Thus, one may store this information in the form of a cell. In our case, YZ is a cell array, whose element $YZ\{J\}$ contains a matrix with the coordinates of the points for the J-th line of magnetic field (in our example, to a full number of 12):

$$YZ\{J\} = \begin{bmatrix} y_J(1) & z_J(1) \\ y_J(2) & z_J(2) \\ \cdot & \cdot \\ \cdot & \cdot \\ \cdot & \cdot \\ y_J(n_J) & z_J(n_J) \end{bmatrix} \tag{6.1}$$

Matrix manipulation and arithmetics
There are plenty of matrix manipulation operations that are performed with extreme efficiency by using internal functions and operators. In table 6.1 we list a number of them, suggested for initial practice.

We recall that the above defined operator ':' deserves a special mention. It gives high flexibility to the syntax. By now, we suggest to check that M(:, :) = M(1: end, 1: end) = M and that M(:) gives the elements of M in array form as a column vector.

Consider table 6.2 for some practice with matrix algebra.

As a special feature, we call the readers' attention to the possibility of performing global, as well as *element-by-element* operations with matrices. The latter are very

Table 6.1. Some matrix manipulation operations.

Command	Action
>> a = [1 2 3]	Define an array (row vector)
>> b = [2 3 1]	Define an array (row vector)
>> c = [3 1 2]	Define array (row vector)
>> M=[a;b;c]	Define matrix by vertical concatenation
>> M(2,:)	Select row 2 of M
>> M(1,end)	Last element of row 1 of M
>> diag(M)	Diagonal elements of M
>> M.'	Transpose of M (M^T in standard notation)
>> M'	Conjugate transpose of M
>> sum(diag(M))	Trace of M
>> trace(M)	Trace of M
>> sum(M(M==1))	Sum over all elements of M equal to 1
>> IM=inv(M)	Inverse matrix of M
>> size(M)	Obtain dimensions of M
>> ones(2,5)	Define a matrix with 1's of size 2×5
>> zeros(5,2)	Define a matrix with 0's of size 5×2
>> eye(5)	Define identity matrix of size 5×5

Table 6.2. Some operations of matrix algebra. One may use the matrix M defined in table 6.1 for some practice.

Command	Action
>> N = M+1	Addition (implicitly of identity)
>> M+N	Matrix addition
>> M-N	Matrix subtraction
>> M*N	Matrix multiplication
>> M^2	Matrix power
>> M/N	Right divide MN^{-1}
>> M\N	Left divide $M^{-1}N$
>> M.*M	Multiply element by element
>> M.^2	Power element by element
>> M./N	Divide element by element
>> M.\N	Left divide element by element

useful in the evaluation of fields as will be seen in the forthcoming example. Just recall that the computation of electromagnetic fields involves operations with physical quantities at the same points of some region of space. Using matrices, whose elements relate to the position in space will be a useful representation.

There is much more concerning the mathematical operations that one may perform with matrices. We will just mention two points of interest for some applications that are extensively documented on the web.

- Relational operations are accessible through the operators ~ (negation), & (and), || (or), == (equal), > greater than, >= (greater or equal), < (less than), <= (less or equal), ~= (unequal). One may figure out how this works by using simple examples. For instance, based on the matrix M defined in table 6.1 one may check that >> M(M==1)=0 replaces 1's by 0's, that >> M(M==1 | M==2)=0 has the same result for 1's and 2's, and that >> length(M(M==1)) counts the number of occurrences of the value 1.
- A good number of mathematical resources is included through the built-in functions in the distribution, ranging from basic trigonometric operations to special functions as Bessel functions, elliptic integrals, etc. Some examples of interest for our purposes are given in section 6.1.6. Additional functions, mathematical algorithms and methods may be downloaded from https://www.mathworks.com/matlabcentral/fileexchange

6.1.3 Graphics

Graphical representation is one of the main achievements of MATLAB. As said, a very flexible interface allows to modify the features of a plot that was stored in a previously created figure object. Below, we review some basic operations for creating and modifying figures.

To start with, we suggest to practice with the basic operation, i.e. 2D plots for vector data. For instance, one may use the following sequence

```
>> x=-2:0.01:2;
>> y=x.^4-x.^2+x-1;
>> plot(x,y)
```

The function plot() creates a figure window and produces a basic plot of the function $y = x^4 - x^2 + x - 1$. Next, one may change formats through the extensive list of options in the menu: change line style, include labels, insert comments, formulas, etc, and eventually save the figure or export to a number of additional formats so as to share it with other programs.

Example

Let us now consider a more sophisticated graphical output. As illustrated in figure 6.2, we propose to plot the magnetic field lines of a dipole source with arbitrary orientation. For this, we will put forward a script that uses some of the above features of the language in order to evaluate $\mathbf{B(r)}$ at some region of space ($\mathbf{r} = (x, y, z)$) and later use the graphical functions to display such lines. We will evaluate the well known expression

$$\mathbf{B(r)} = \frac{\mu_0}{4\pi}\left[\frac{3\mathbf{r'}(\boldsymbol{\mu} \cdot \mathbf{r'})}{r'^5} - \frac{\boldsymbol{\mu}}{r'^3}\right] \qquad (6.2)$$

with $\mathbf{r'}$ being the relative position respect to the magnetic dipole ($\mathbf{r'} = \mathbf{r} - \mathbf{r}_M$). First, we need to define a grid of points (x, y, z), that covers the region of interest, perform the mathematical operations in the above expression, and plot the lines of **B** for a given position and configuration of the dipole. As seen below, the use of matrix arithmetics and built-in graphical functions are very useful for this task.

To start with, we introduce the function meshgrid(). In order to figure out how it works, just type

```
>> y=1:4 ; z=-2:2;
>> [Y,Z]=meshgrid(y,z);
```

It is apparent that one has created a rectangular grid of points that cover the region defined by: $1 \leqslant y \leqslant 4$, $-2 \leqslant z \leqslant 2$.[8] The information is stored in the matrices Y and Z. Notice that the rows of Y replicate the vector y and that the columns of Z replicate the vector z, such that $[Y(i, j), Z(i, j)]$ gives the coordinates of the point $[y(j), z(i)]$. These matrices may be used to evaluate functions of two variables in the region of interest. In particular, one may codify equation (6.2) by the commands included in the script *dipole_field.m*.

```
%%%%%%%%%%%%%%%%%%%%%%%%%%%%%%%%%%%%%%%%%%%%%%%%%%%%%%%%%%%%%%%%%%%%%
%                     File:  dipole_field.m
%
% This script evaluates Eq.(6.2) for a dipole at the point (0,0,0)
% wihtin the plane x=0. It uses the magnetic moment (0,my,mz)
%
% The region of space corresponds to -6< y <6 ; -6< z <6
%
%%%%%%%%%%%%%%%%%%%%%%%%%%%%%%%%%%%%%%%%%%%%%%%%%%%%%%%%%%%%%%%%%%%%%
%
[Y,Z]=meshgrid(-6.0:0.05:6.0,6:-0.05:-6); % grid of points
%
%
By=3*Y.*(my*Y+mz*Z)./(Y.^2+Z.^2).^(5/2) ...      % y-component
                   -my./(Y.^2+Z.^2).^(3/2);

Bz=3*Z.*(my*Y+mz*Z)./(Y.^2+Z.^2).^(5/2) ...      % z-component
                   -mz./(Y.^2+Z.^2).^(3/2);
```

Then, having evaluated the field components at the points of grid, and loaded in the matrices B_y and B_z, one may plot the vector field lines. For this purpose, we use the internal function streamline(). As one may see in the code accompanying the figure, this function inputs the grid points $[Y, Z]$, the vector field at such points $[B_y, B_z]$, and the arrays *starty* and *startz*. These arrays are just the coordinates of the 'starting points' to be used by the function for evaluating the field lines. Thus, for a given starting point, the program plots the line generated by integration from such point (see equation (12.6) to figure out how this works).

Two final remarks about the current example must be done. First, we notice that the magnetic field lines have been oriented using arrow heads. They are produced with help of the function arrow() introduced in section 6.1.1. Thus, defining

---

[8] 3D grids may be created by using the syntax >> meshgrid(x,y,z).

general purpose functions may be profitable for further use. Second, we recall that, once more, matrix algebra is of much help. Having solved the problem for the vertical orientation of the magnetic dipole, one may obtain a rotated solution just by using the above defined rotation matrix R (section 6.1.2) and straightforward matrix multiplication:

$$\begin{bmatrix} B_y^r \\ B_z^r \end{bmatrix} = \begin{bmatrix} \cos\theta & \sin\theta \\ -\sin\theta & \cos\theta \end{bmatrix} \begin{bmatrix} B_y \\ B_z \end{bmatrix} \rightarrow \text{>>Byzr(i) = R*[By(i); Bz(i)]}$$

where i represents the i-th point of the grid[9].

Some other very useful graphical functions are: `contour()` that plots the contour levels of a 2D function $f(x, y)$, and `surf()` that builds the 3D plot $z = f(x, y)$. They will appear in some of our applications.

6.1.4 Mathematical analysis tools

The availability of mathematical resources in MATLAB is high and increasing. Just to give an idea, below we mention some of them, that will be of interest for further use in part III of this book.

Computational geometry
MATLAB allows to deal with geometrical objects, work with regions, boundaries, areas, etc. An elementary use of such capabilities is done in chapter 9 (see figure 9.5). There, we take advantage of such pre-defined functions as `polyshape()`, `intersect()` and `area()` that efficiently define polygonal regions, evaluate their intersections and areas.

Numerical integration
A number of built-in functions are available with special methods for different specific conditions, including improper integrals, complex line integration and multidimensional cases. For further reference we want to mention `trapz()`, `integral()`, `integral2()`, `integral3()` that will be extensively used in our applications.

Optimisation
Numerical optimisation is a highly relevant technique in the statement of many physical problems. In particular, this affects a number of phenomena in part III[10]. A rather efficient and flexible possibility for solving such problems is to use the functions included in the MATLAB optimisation toolbox. The general optimisation problem, that may be solved with the function `fmincon()` reads

---

[9] As said before, when a matrix is designed in the form $M(:)$ it is represented as a column vector with concatenated successive rows: $M(:) = [M(:, 1); M(:, 2); ...;M(:, \text{end})]$.
[10] Recall that variational principles are an extended practice in physical theories.

$$\min_{X} F(X) \text{ such that } \begin{cases} C_{INEQ}(X) \leqslant 0 \\ C_{EQ}(X) = 0 \\ A_{INEQ}X \leqslant B_{INEQ} \\ A_{EQ}X = B_{EQ} \\ L_{BOUND} \leqslant X \leqslant U_{BOUND} \end{cases} \tag{6.3}$$

That is to say, we want to minimise the function $F(X)$ of the variable X (possibly an array of variables), under a number of conditions. Specifically, one may have non-linear inequality and equality conditions, linear inequality and equality conditions, and also the restriction that X belongs to some bounded region. Thus, we notice that in the statement (6.3) F is a vector valued function that returns a scalar, C_{INEQ}, C_{EQ} are vector valued functions that return vectors, A_{INEQ}, A_{EQ} are matrices, and B_{INEQ}, B_{EQ}, L_{BOUND}, U_{BOUND} are vector quantities.

In our case, the function F will be quadratic in many instances, i.e.

$$F(X) = X^T M X + f^T X \tag{6.4}$$

with M being a square matrix and f^T the transpose of a constant vector, and it will be possible to use specialised optimisation algorithms as those in the function `quadprog()`.

In order to see how all this works, we present an example related to the problem of steady transport current in the electromagnetic theory. As sketched in figure 6.3, we propose to solve how the electrical current density distributes within a non-homogeneous conducting medium shaped in the form of a long tape.

A full transport current I_{tr} is fed along the y-axis of the tape, that occupies the region $-a \leqslant x \leqslant a$, $|y| < \infty$, $|z| < d \approx 0$. The material has a non-homogeneous resistivity given by

$$\rho(x) = \rho_0(1 - c|x|) \tag{6.5}$$

and we want to find the current density distribution across the width of the tape, i.e. $J(x)$. The problem may be formulated in terms of the variational statement exposed in section 1.2.2[11]. Specifically, one may use the principle of minimum entropy production (exercise R1-3) and use the dissipation function in equation (1.30) in the one dimensional form. We end up with the statement

$$\min_{J(x)} \int_0^a \frac{1}{2}\rho_0(1 - c|x|)J(x)^2 \, dx \quad \text{such that} \quad 2d\int_{-a}^a J(x)dx = I_{tr} \tag{6.6}$$

i.e. one must find the function $J(x)$ such that the dissipation of energy per unit volume and per unit time is minimised, subject to the overall transport current

---

[11] Alternatively, this problem may also be solved by recalling the concept of parallel resistors. This method may be used as a double-check of the optimisation procedure explained here.

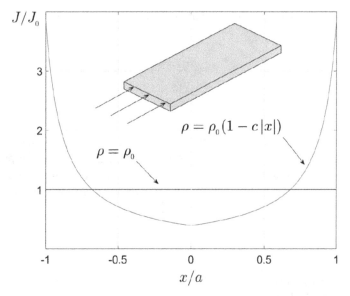

Figure 6.3. Calculation of the current density distribution in a conducting tape with either constant resistivity ($c = 0$ in equation (6.5)) or inhomogeneous resistivity ($c = 0.9$ in equation (6.5)) by application of equation (6.7), a particular case of (6.4). Units are defined by the full transport current ($J_0 \equiv I_{tr}/4ad$) and the half width of the sample, a 🔗. Source codes available at https://iopscience.iop.org/book/978-0-7503-2711-4.

condition. This principle may be transformed into the statement in equation (6.3) (actually into its particular version (6.4)), just by identifying $J(x)$ at a set of points x_i across the width of the tape, with the unknown vector X. Specifically, equation (6.6) takes the form[12]

$$\min_{\mathbf{J}} \mathbf{J}^T \mathbf{P} \mathbf{J} \quad \text{such that} \quad \mathbf{A}_{EQ} \mathbf{J} = \mathbf{I}_t \tag{6.7}$$

In our MATLAB code, this statement is solved by typing the command

```
J=quadprog(P,[],[],[],Aeq,It,Lb,Ub)
```

Notice that the components of expression (6.7) are:
   J: column vector with the current density at a collection of points across the width of the sample $\{x_i, i = 1, \ldots, N\}$, i.e. $\mathbf{J}(i) = J(x_i)$
   P: diagonal resistivity matrix, such that $\mathbf{P}(i, i) = (1 - c|x_i|)$
   $\mathbf{A}_{EQ}$: row vector defined by the command >> Aeq = ones(1,N)
   $\mathbf{I}_t$: scalar of value I_{tr}

---

[12] We notice that global multiplying constants in the expression that is minimised are irrelevant.

The set of points that discretise the half width of the sample (recall the symmetry) are defined by

```
>> x=(1:N)/N
```

and the matrix **P**

```
>> P=diag(1-c*x)
```

As shown in figure 6.3, by changing the value of the parameter c, one may investigate the effect of having a non-homogeneous resistivity $\rho(x)$. In particular, we emphasise the comparison to the case $\rho(x) = \rho_0$ which reproduces the well known property that, for homogeneous resistivity, the resulting current density distributes uniformly.

6.2 GNU Octave

OCTAVE is a scientific programming language that takes part of the GNU project[13] for free software. Full information, download and installation instructions may be found at https://www.octave.org. Its conception and capabilities are very close to those of MATLAB and the portability between both languages is large and increasing. In this regard, the above presentation concerning the user's interface, language and tools is perfectly valid for those who initiate their experience with OCTAVE.

Although a stand-alone installation of OCTAVE does not require either a high performance computer or much space of disk, users may also benefit of a web interface that runs on the browser and may perform a number of operations on scripts. It may be accessed in a very flexible way from https://octave-online.net and is very advantageous for collaborative work.

Concerning the software provided in this book, most of it was developed in MATLAB and later tested in OCTAVE. In fact, in our case portability is nearly complete. Just in several instances, we include comments on the alternative formulation in both languages. It must be said, however, that while writing this book, a new release appeared that, among other things, increases the compatibility of both languages. Nevertheless, as far as we have been able to understand, the language resident functions seem more efficient than those created for full compatibility. In particular, this affects to the *Optimization* package. For instance, related to the general non-linear optimisation problem stated in (6.3), we have preferred to use the 'conventional' solver sqp() than the recently incorporated function fmincon(), and the 'conventional' quadratic solver qp() than the recently incorporated function quadprog() that provide compatibility with MATLAB. Related to this, below we give an example of the simple operation that allows to switch between one version and the other in those cases for which compatibility is not complete. Simply, by

---

[13] http://gnu.org.

commenting/uncommenting some lines of the code, the script will either run in one program or the other:

```
%%%%%%%-------------        MATLAB BLINDED    -----------%%%%%%%
%X= quadprog(M,f,Aineq,bineq,Aeq,beq,Lb,Ub,x0);
%%%%%%%-------------        OCTAVE USERS    -----------%%%%%%%
%
X=qp(x0,M,f,Aeq,beq,Lb,Ub,ALb,Aineq,AUb,Lb,Ub);
%
%%%%%%%-------------------------------------------------%%%%%%%
```

```
%%%%%%%-------------        MATLAB USERS    -----------%%%%%%%
%
X= quadprog(M,f,Aineq,bineq,Aeq,beq,Lb,Ub,x0);
%
%%%%%%%-------------        OCTAVE BLINDED    -----------%%%%%%%
%X=qp(x0,M,f,Aeq,beq,Lb,Ub,ALb,Aineq,AUb,Lb,Ub);
%%%%%%%%-------------------------------------------------%%%%%%%
```

In any case, small changes in the syntax must be advised. For instance, in the above fragments of the script related to figure 6.3 one may notice that the syntax of OCTAVE's qp() is slightly different to that of MATLAB's quadprog(). Also, the statement of the problem changes a bit. Thus, in the quadratic problem, MATLAB's expression

$$A_{INEQ}X \leqslant B_{INEQ} \tag{6.8}$$

is replaced by OCTAVE's

$$A_{Ub} \leqslant A_{INEQ}X \leqslant A_{Lb} \tag{6.9}$$

Along the book, when no comment is made, we mean that the script has been tested and works in both languages.

Part III

Applications and utilities

Part III

Chapter 7

The resistive transition

As emphasised in chapter 4, resistive measurements $R(T)$ have been a widespread experimental technique, not only for giving a straightforward characterisation method of superconducting materials. In particular, combined with the response to an applied magnetic field, i.e. when the so-called *magnetoresistance* $R(H_a, T)$ is measured, experiments may reveal much about the underlying physical mechanisms. Among other aspects, one obtains information about the dynamics of vortices and its influence on the superconducting transition. Below, we describe a specially relevant effect, the so-called broadening of the resistive transition that captured the attention of many researchers soon after the discovery of high temperature super-conductivity. A series of utilities for evaluating the related phenomena and analysing the experimental results will be provided throughout this chapter.

7.1 The broadening of the resistive transition

In the late 1980s intense research activity was devoted to the nature of the superconducting transition of cuprates. It was recognised early on that the extremely small value of the coherence length ξ of these materials, together with the 'high temperatures' involved in many experiments, seriously enhance the role of thermally activated phenomena. Noticeably, this fact may have strong implications as concerns the interpretation of experiments. In particular, Yeshurun and Malozemoff [1] emphasised that the determination of critical parameters as T_c and H_{c2} may be noticeably hampered by this. In appropriate terms, instead of obtaining the $H_{c2}(T_c)$ line, one may be measuring the so-called 'irreversibility line'. This threshold is determined by the condition that the vortex lattice 'melts' and can no longer be pinned by whatever the pinning landscape is present. What is relevant in cuprate superconductors is that these effects extend over a well observable range of applied fields and temperatures. In particular, as shown by Tinkham [2], this concept may well explain the observation that the resistive transition of YBCO samples noticeably broadens in the presence of a magnetic field. Physically, the $R(T)$

curves show a rather smooth transition between zero resistance and the normal state. The intermediate region corresponds to a resistive yet thermodynamically[1] super-conducting state associated to activated flow of vortices.

Let us see how one may deal with the above situation so as to get some information from the experimental observations and analyse the underlying physical mechanisms. First, considering that the problem may be parameterised by the ratio between the pinning potential and thermal energies, Tinkham introduced a normalised barrier height for the activation of vortices. It is given by

$$\gamma = \frac{(1 - t)^{3/2}}{b} \tag{7.1}$$

In this expression, one uses the reduced temperature $t \equiv T/T_c$ and assumes a two-fluid model dependence on T (see section 2.1). b is a normalised magnetic flux density $b \equiv B/B_0$ in terms of the characteristic value (in S.I. units)

$$B_0 \equiv \frac{6\sqrt{3}\,\Phi_0^2}{k_B T_c} J_{c_0} \tag{7.2}$$

with J_{c_0} being the material's critical current density at zero field and temperature, Φ_0 the flux quantum, k_B the Boltzmann constant, and T_c the critical temperature. In passing, we note that in this chapter we will use $\mathbf{B} \approx \mu_0 \mathbf{H}$ because self fields may be neglected.

For a deeper discussion of equation (7.1) the reader is addressed to the work by Tinkham [2, 3]. Here, we just interpret its physical meaning. On the one side, it is apparent that the activation energy goes to zero as T goes to T_c. The specific power law is related to the underlying physical mechanism. On the other side, it is of notice that a high value of B_0 means that one may apply strong magnetic field with non-significant effects due to barrier reduction as temperature increases. Now, in order to connect these ideas with the experimental facts, Tinkham took advantage of a well known work by Ambegaokar and Halperin [4] where these authors analysed the appearance of voltage across a Josephson junction as related to thermal noise. He argued that the dynamic of vortices over the complex pinning landscape in a macroscopic sample is analogous to the heavily damped case studied in the paper. Thus, he proposed to use the low current limit[2] expression from that article

$$r \equiv \frac{R}{R_n} = [I_0(\gamma/2)]^{-2} \tag{7.3}$$

with R_n being the normal state resistivity and I_0 the modified Bessel function.

We recall that the above expression has been tested against experimental data with a good degree of consistency. In fact, by using B_0 as a single fit parameter, $R(T)$ curves for applied magnetic induction in the range 0–9 T could be reproduced. As a satisfactory test for the model, the values of J_{c_0} deduced from equations (7.3) and (7.2) were well within expectations for the samples under consideration

---

[1] The superconducting order parameter is different from zero though resistance is finite.
[2] The low current approximation usually accommodates the experimental conditions in resistivity measurements. Later we will introduce a method that overcomes this simplification.

For nearly three decades, the above model has been used with success either in its original form or modified for specific purposes.

We call the readers' attention to, upon having data on the physical parameters T_c and J_{c_0}, one may anticipate the relevance of the resistive broadening for a given sample.

7.2 Evaluation of resistance and activation energies

Below, we will concentrate on the properties of the resistive transition modelled by equation (7.3) with emphasis on the physical parameters T_c and B_0. The analysis will be extended to the implementation of a method that allows to 'invert' the relation and straightforwardly obtain the activation energy in terms of the (measurable) resistance, i.e.

$$\gamma = 2I_0^{-1}[r^{-1/2}] \tag{7.4}$$

Here, I_0^{-1} stands for the inverse of the modified Bessel function and $r^{-1/2}$ for the multiplicative inverse of the square root of r. Notice that, using this procedure one is neither constrained to assume a given parametric dependence for $\gamma(T)$ to be inserted in the direct function $R(T)$ nor to subsequent fitting procedures. Our methodology relies on some simple MATLAB codes listed below. With these, the reader may straightforwardly evaluate either (i) the broadening of the resistive transition $r(t)$ as a function of the applied magnetic field[3] or (ii) predict the magnetoresistance in isothermal conditions as a function of the reduced temperature T/T_c. Typical results are shown in figures 7.1 and 7.2. The corresponding source codes are listed at the end of the book and named after

- R_T.m (main) ; T_resistivity.m ; createfigure_T.m
- R_B.m (main); B_resistivity.m ; createfigure_B.m

From a different point of view, one could also be interested in straightforwardly deriving activation energies $\gamma(T)$ from available resistivity data $R(T)$. Below we show results of the inversion procedure based on the application of equation (7.4). The 'inversion' of $R(T)$ measurements may be implemented through the following source codes

- R_data_inv.m (main); T_resistivity3.m ; createfigure_re-cover.m

In the following example, the 'original' data (inset of figure 7.3) are simulated through equation (7.3) and plotted aside.

7.3 Extensions

The above treatment of the resistive transition has simplified two aspects that could be of interest in some applications. First, we have mentioned that equation (7.3) is a 'low current' limit, and a quantitative statement regarding the validity of the

---

[3] In these experiments one may use $B \approx \mu_0 H_a$. The applied magnetic field is dominant.

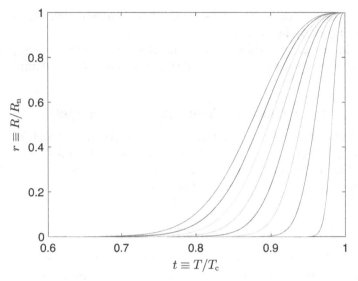

Figure 7.1. Normalised resistance versus reduced temperature as calculated with equation (7.3) for different values of the applied magnetic induction. From right to left: $B/B_0 = 0.001: 0.0025: 0.02$ 🔗. Source codes available at https://iopscience.iop.org/book/978-0-7503-2711-4.

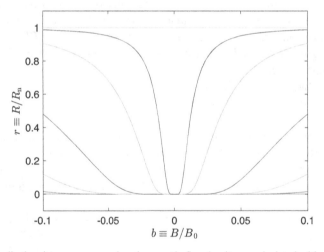

Figure 7.2. Normalised resistance versus reduced magnetic flux density as calculated with equation (7.4) for different values of temperature. From bottom to top: $T/T_c = 0$; 0.2; 0.4; 0.6; 0.8; 0.9; 1 🔗. Source codes available at https://iopscience.iop.org/book/978-0-7503-2711-4.

approximation is awaiting. Also, considering the utility of resistive measurements as a method to reveal the intrinsic activation energy of the samples, we call the readers' attention that we have used 'simulated' $R(T)$ data. As one may want to know how the method performs on data extracted from a real experiment, some additional considerations are needed. These topics will be dealt with within the forthcoming section.

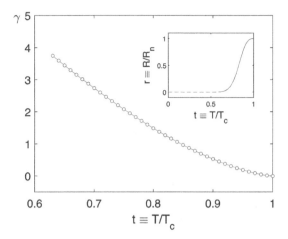

Figure 7.3. Normalised activation energy γ obtained by inversion of the normalised resistance R/R_n curve shown in the inset. The dashed line indicates the lower part of the $r(t)$ curve ($r < 0.01$) that was ignored in the inversion procedure. Source codes available at https://iopscience.iop.org/book/978-0-7503-2711-4.

7.3.1 Influence of the applied current

As said, the ansatz that the resistive transition curve is independent of the current injected along the sample must be reconsidered. In particular, we recall that the 'low current' approximation implies a constant resistivity (independent of I) at given conditions of field and temperature. Thus, notice that the value of the transport current I is not a parameter of the behaviour predicted either for $R(T)$ or $R(B)$. However, as one could expect, non linearities could be important as the current through the sample increases. A closer view to the work by Ambegaokar and Halperin will help us to evaluate the expected voltage drop for any value of I. The main result in their article is the full integral expression that gives the normalised voltage as a function of the applied current density, and on the normalised activation energy.

$$\frac{V}{I_c R_n} = \frac{4\pi}{\gamma}\left\{(e^{\pi\gamma x} - 1)^{-1}\left[\int_0^{2\pi} d\theta\, f(\theta)\right]\left[\int_0^{2\pi} d\theta\, \frac{1}{f(\theta)}\right]\right.$$
$$\left. + \int_0^{2\pi} d\theta\, f(\theta)\int_\theta^{2\pi} d\alpha\, \frac{1}{f(\alpha)}\right\}^{-1} \tag{7.5}$$

Here I_c represents the critical current of the system, the applied current is given in reduced units $x \equiv I/I_c$, and the function $f(\theta)$ is defined by

$$f(\theta) = e^{\gamma(x\theta + \cos\theta)/2} \tag{7.6}$$

We recall that the second term in equation (7.5) contains a multiple integral with the lower limit as an independent variable. Two possibilities will be described on how to cope with this for numerical evaluations.

A set of $V(I)$ curves obtained with the help of equation (7.5) are shown in figure 7.4. From left to right, they correspond to increasing values of the normalised activation energy γ.

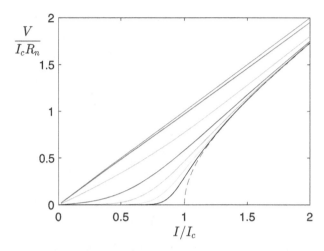

Figure 7.4. Normalised voltage current curves for different values of the activation energy as calculated with equation (7.5). From left to right $\gamma = 0.01, 0.5, 2, 5, 10, 20, 40$. The dashed line corresponds to the limit $\gamma \to \infty$ (see text) $\mathcal{O}$. Source codes available at https://iopscience.iop.org/book/978-0-7503-2711-4.

Notice that the full range $0 < \gamma < \infty$ is bounded by two simple curves. On the one side (as expected):

$$V \approx IR_{\mathrm{n}} \qquad (\gamma \to 0) \tag{7.7}$$

i.e. in the presence of a negligible barrier height as compared to the thermal energy, the transport current drives the sample into the resistive state.

On the other side:

$$\frac{V}{I_c R_{\mathrm{n}}} \approx \begin{cases} 0, & x \leqslant 1 \\ \\ \sqrt{x^2 - 1}, & x > 1 \end{cases} \qquad (\gamma \to \infty) \tag{7.8}$$

i.e. a very high barrier implies a practically zero voltage for $I < I_c$ and $V \approx IR_{\mathrm{n}}$ when I increases well above I_c, i.e..: after some transition region.

Concerning the source codes related to figure 7.4 they are available and named after

- `VI_curves_AH.m (main); f_AH.m; inv_f_AH.m; int_inv_f_AH.m`
- `interp_int_inv.m; int_combined.m; createfigure_V_I.m`

It is important to notice that the numerical integration performed by the above scripts uses a standard quadrature method that loses accuracy as γ increases. In practice, one may use them confidently for $\gamma \leqslant 30$. For the interested reader, some possible alternative is discussed further in the proposed exercises[4].

----

[4] The $V(I)$ curve corresponding to $\gamma = 40$ in figure 7.4 was obtained by such method.

To finish this section, we recall once again that, strictly speaking, the theory by Ambegaokar and Halperin applies to a single overdamped Josephson junction. However, from a phenomenological point of view, the essential ideas can be applied to macroscopic samples in which voltage is induced by flux depinning across a basically homogeneous network of pinning sites.

7.3.2 Activation energies from experimental $R(T)$ data

In section 7.2 we have provided an example of numerical inversion of $R(T)$ data so as to extract the underlying activation barrier for vortex depinning. Certainly a plausible temperature dependence $\gamma(T)$ was assumed in that case. Nevertheless, it could well be the purpose of the reader to obtain this behaviour for a given sample whose actual resistivity has been measured at the laboratory. Let us show how this works with real data. In figure 7.5 we plot the resulting inversion for measurements in a thin film of the high T_c superconductor YBaCuO (2016, private communication).

From the physical point of view, a straightforward conclusion of the result is that a single power law does not suffice to describe the behaviour of the sample close to T_c. In fact, as revealed by the logarithmic plot, the curve shows two well-defined regimes. At the highest temperatures, a linear fit to the data (red points) indicates a power law with exponent $n = 0.66$. On the other hand, at lower temperatures, a power law with $n = 2.57$ is obtained. According to the literature [2] such a difference in behaviour is indeed expected if one considers that close to T_c (smaller exponent)

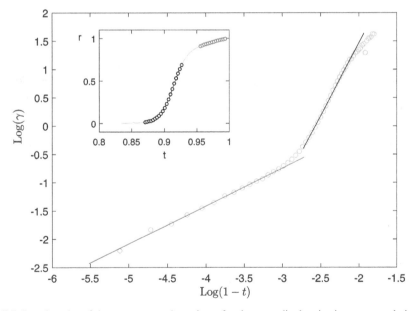

Figure 7.5. Log–log plot of the temperature dependence for the normalised activation energy γ derived for a thin film of YBCO. The inset shows the original $r(t)$ data recorded under an applied field of 0.1 T perpendicular to the film. Well-defined power law dependencies are emphasised by the linear fits in the logarithmic plot 🔗. Source codes available at https://iopscience.iop.org/book/978-0-7503-2711-4.

vortex depinning is irrelevant, and the activation energy is less significant. Vortices are said to form a 'liquid phase' above the so-called 'melting temperature' T_m. On the other hand, one expects higher values of the exponent, when describing the landscape of vortices hopping along potential barriers below T_m.

The software for analysing this kind of situation is provided in the following scripts:

- get_data_r.m ; R_data_inv_r.m (main); inverter_r.m
- createfigure_recover_r.m

Together with them, a file including the experimental data (R_YBCO.dat) is included for practice.

7.4 Review exercises and challenges

7.4.1 Review exercises

Review exercise R7-1
Study the effect of changing the value of the applied field B in the $R(T)$ curves (corresponding to equation (7.2)). Replot figure 7.1 for $B/B_0 = 0.01: 0.025: 0.2$. Identify the different curves

Hint: modify the script R_T.m.

Review exercise R7-2
Suppose that figure 7.1 corresponds to resistance measurements for a sample of YBaCuO. Assume a typical value for the material parameters $T_c = 95K$ and $J_{c_0} = 10^6$ A/cm^2. Is the curve for $B/B_0 = 0.02$ reasonably to be expected in a low temperature lab? What would be the maximum broadening to be expected in a lab with a 14T magnet? Make a plot of the corresponding $r(t)$ curve.

Review exercise R7-3
Study the effect of temperature in the magnetoresistivity as described in figure 7.2. Assuming that equation (7.3) remains valid up to T_c analyse the evolution $T/T_c = 0.9: 0.01: 1$. Is this realistic? If true, what would this imply for the behaviour of $R(H \rightarrow 0, T)$?

For a possible (empirical) bypass to this problem, the reader is addressed to the paper by Tinkham [2], where equation (7.3) is utilised to obtain predictions as relative to the master curve for $B = 0$.

Hint: modify the script R_B.m.

Review exercise R7-4
Related to figure 7.3, modify the script R_data_inv.m as well as the function
T_resistivity3() so as to plot the 'recovered' curve $\gamma(T)$ for a set of values of
the applied magnetic field. For example, we suggest that the reader takes the values
B/B_0 = 0.005, 0.01, 0.02, 0.03. Use a log–log plot to emphasise the power law $(1 - t)^n$
assumed in the simulation of the $R(T)$ data. For visualisation you may also need to
modify the script create figure_recover.m.

Hint: it is useful to enter B as an argument of the function T_resistivity3().

Review exercise R7-5
Repeat the previous evaluation (review exercise R7_4), but in this case take a given
value of B/B_0 (for example 0.03) and investigate the effect of changes in the power law,
i.e. use n = 1, 1.5, 2, 3. These values are not unreasonable as related to a number of
physical mechanisms. Draw conclusions on the shape of the $R(T)$ curves in terms of n.

Hint: it is useful to enter n as an argument of the function T_resistivity3().

Review exercise R7-6
Study the validity of the 'low current' approximation

$$\frac{V}{I_c R_n} \approx [I_0(\gamma/2)]^{-2}\frac{I}{I_c} \tag{7.9}$$

widespread used in resistivity measurements. This linear approximation means that the
experimental resistance does not depend on the applied transport current.

Hint: for the above purpose, one may use the scripts related to figure 7.4 and
show, for instance, that with $\gamma \leqslant 1$ the exact dependence obtained from equation
(7.5) is well replaced by the linear approximation. A suggested set of values is
$0 < I/I_c < 2$ and γ = 0.1, 0.5, 1.2.

7.4.2 Challenges

Exercise C7-1
As said, the integral representation of the voltage current characteristic in equation
(7.5) may be hard to deal with from a numerical point of view. In particular, standard
quadrature methods as those used by the MATLAB function integrate() fail to
converge to the correct values in some circumstances ($\gamma \gg 1$). The reader is invited to
check this by straightforwardly comparing the result for $\gamma = 40$ to the approximation
for high barriers (equation (7.8)). A possible solution for this handicap is to use another
'integrator'. For instance, one may take advantage of the built-in function

`NIntegrate[]` in MATHEMATICA. Use this or a choice of your own to verify that the results for γ converge to the correct solution. Also, this may be a way to verify the previous calculations for moderate values of γ.

Hint: in case of using the function `NIntegrate[]` one has to be advised that some issue could arise in MATHEMATICA. This relates to the fact that one cannot make a symbolic evaluation within the numeric integration (as required by defining a function through a limit of integration). This problem may be fixed by using the function `NumericQ` that avoids an early evaluation, prior to having numerical values.

Exercise C7-2

Throughout this chapter, we have worked out a number of calculations related to the resistivity of type-II superconductors in dimensionless units. Now we propose to go quantitative, based on the reproduction of figure 1 in [2]. First, one must derive the expression for $\Delta_i \equiv T_i - T_c$ as defined in that paper, i.e. for a given level of resistance $r_i \equiv R/R_n$ one must obtain the temperature of occurrence of such level of resistance in a given level of magnetic field $t_i(r_i, b)$ (in dimensionless units). This may be done by a combination of equations (7.2) and (7.3). Then, by input of the parameter values T_c and J_{c_0} one has the actual temperature shift Δ_i (in Kelvin) that corresponds to each value r_i of the set and for any value of B. Eventually, taking the $B = 0$ curve as a reference, one plots the displacement Δ_i relative to this curve[5], just to check the impressive reproduction of the experimental facts achieved by the model.

Hint: the inversion procedure introduced to generate figure 7.3 may be straight-forwardly modified for the above purpose.

References

[1] Yeshurun Y and Malozemoff A P 1988 Giant flux creep and irreversibility in an Y-Ba-Cu-O crystal: an alternative to the superconducting-glass model *Phys. Rev. Lett.* **60** 2202

[2] Tinkham M 1988 Resistive transition of high-temperature superconductors *Phys. Rev. Lett.* **61** 1658

[3] Tinkham M 1996 Introduction to Superconductivity *International Series in Pure and Applied Physics* (New York: McGraw-Hill)

[4] Ambegaokar V and Halperin B I 1969 Voltage due to thermal noise in the dc Josephson effect *Phys. Rev. Lett.* **22** 1364

---

[5] The physical background of this manipulation is clear: according to the simplified (but rather realistic) model used by Tinkham, the $B = 0$ curve should be a step function. As reality is a bit more complex, one may 'counter-balance' the additional effects by plotting relative to this.

IOP Publishing

Macroscopic Superconducting Phenomena
An interactive guide
Antonio Badía-Majós

Chapter 8

Flux transport in type-II superconductors

As emphasised in chapter 4 the transport of magnetic flux, either penetrating in or abandoning a type-II material, deserves much attention in macroscopic superconductivity. On the one hand, a large number of experimental studies recording the response of the sample to a variety of magnetic excitation processes give us detailed information on underlying physics. In particular, this involves electromagnetic cycling processes that tell much about vortex pinning mechanisms at different regimes of field and temperature. On the other hand, as shown in section 4.2, this connects to the very relevant macroscopic parameter J_c, i.e. the material's critical current density that is the core of the phenomenological critical state model. This model and its extensions have been the basis for hundreds of applied investigations such as the development of superconducting cables, magnets and levitation systems.

Here, we set the base for the quantitative analysis of experiments and applications that may be modelled in the realm of the critical state model and some related statements. A number of specific problems whose treatment builds upon these principles will be considered in detail in ensuing chapters.

We will concentrate on the upgrade of the simplest version of the critical state model (issued in section 3.3, i.e. Bean's model applied to trivial geometry). Two basic points of interest will be focused on. First, the inclusion of finite-size effects, impossible to skip in practical situations, will be a feature of this chapter's resources. Also, the (implicit) infinite resistivity ansatz will be bypassed, so as to include possible time relaxation effects that may be of importance for certain experimental conditions. Along these lines, we will supply source codes dealing with a generalised variational statement of the critical state model applied to (i) transport and magnetisation problems in 1D and 2D geometries, (ii) the response of superconducting samples to non-uniform magnetic fields, (iii) the so-called *piece-wise* approximation to deal with resistivity effects and (iv) the celebrated *power-law* model, which is the basis of the vast majority of calculations in engineering applications[1].

8.1 The penetration of magnetic fields in superconductors

Recalling section 1.2.4, at the macroscopic level, the penetration of magnetic fields in conducting media may be formulated through an evolutionary variational statement that updates the state of the system along successive time steps. As said, this approach offers a very convenient alternative to the conventional differential equation form, because, with the exception of the case of ohmic conductors, the diffusive penetration process will be non-linear and generally this leads to a tough mathematical problem.

8.1.1 Variational statement

Let us briefly recall that the *time advancing* solution of the Maxwell equations for the electromagnetic field penetrating within conductors may be obtained through a step-by-step minimisation of a functional that measures variations of energy. These variations include internal (recoverable) storage, as well as energy losses. The former stands for the storage of energy in the form of circulating currents, and the latter will be encoded through the dissipation function $\mathcal{P}$. In practice [1], the validity of this approach encloses a wide class of $\mathbf{E}(\mathbf{J})$ laws. Some of them, the most popular in the field of superconductivity, will be considered in the examples treated below.

For convenience, in what follows we consider the *functional* expressed by $\mathcal{F}[\mathbf{A}(\mathbf{r}), \mathbf{J}(\mathbf{r})]$. In general, the role of the vector potential will be useful for incorporating the action of external sources to the problem. They will be considered 'regulated' by the experimentalist. Thus, we will use the notation $\mathbf{A}_S$ for the corresponding vector potential (recalling the concept of 'source' in thermodynamics). Thus, in terms of these variables, the functional introduced in equation (1.37) may be cast in the form[2]

$$\mathcal{F}[\mathbf{J}_{n+1}(\mathbf{r})] = \frac{\mu_0}{4\pi} \int \int_\Omega \left[\frac{1}{2} \frac{\mathbf{J}_{n+1}(\mathbf{r}) \cdot \mathbf{J}_{n+1}(\mathbf{r}') - 2\mathbf{J}_n(\mathbf{r}) \cdot \mathbf{J}_{n+1}(\mathbf{r}')}{\|\mathbf{r} - \mathbf{r}'\|} \right] dV dV'$$
$$+ \int_\Omega \Delta \mathbf{A}_S(\mathbf{r}) \cdot \mathbf{J}_{n+1}(\mathbf{r}) \, dV + \Delta t \int_\Omega \mathcal{P}[\mathbf{J}_{n+1}(\mathbf{r})] dV \qquad (8.1)$$

Here, the following notation has been used:

Ω: region occupied by the superconductor

$\mathbf{J}_{n+1}(\mathbf{r})$: (unknown) current density at point $\mathbf{r}$ and incremented time $t_n + \Delta t$

$\mathbf{J}_n(\mathbf{r})$: (given) current density at $\mathbf{r}$ and previous time t_n

---

[1] The *piece-wise* approximation and the *power-law* model involve modelling the response of the superconductor based on the functional dependence $V(I)$, i.e. the voltage–current characteristic of the material.

[2] This transformation involves classical electromagnetic manipulations. It was proposed as a guided exercise in chapter 1.

$\Delta \mathbf{A_S}$: variation of the vector potential created by the sources (coils, permanent magnets, etc) along Δt

$\mathcal{P}[\mathbf{J}(\mathbf{r})]$: power dissipated per unit volume at a given point within the superconductor along Δt.

For further application and also to give physical insight, it will be useful to consider each term of the previous functional separately, and thus we define

$$\mathcal{F} \equiv \mathcal{F}_{JJ} + \mathcal{F}_{J0} + \mathcal{F}_{JS} + \Delta t \, \mathcal{W}_{JE} \tag{8.2}$$

with

$\mathcal{F}_{JJ}$: self energy of the evolutionary circulating currents

$\mathcal{F}_{J0}$: interaction energy of the evolutionary currents with a 'frozen' distribution, corresponding to the previous time layer

$\mathcal{F}_{JS}$: interaction energy of the evolutionary currents with the magnetic source

$\Delta t \, \mathcal{W}_{JE}$: energy related to the entropy production due to dissipative mechanisms.

The minimisation of $\mathcal{F}$ in equation (8.1) will be applied to 1D and 2D problems that may be considered as particular cases of two configurations: (i) long rods with arbitrary cross section under translationally symmetric excitation and (ii) circular cylinders under rotationally symmetric fields (see figure 8.1). In both cases, the electromagnetic fields are only a function of coordinates in the plane. Here, we will choose the pair (x, z) in the first case and (ρ, z) in the second one.

These configurations have served as model systems to investigate a number of interesting physical properties. Thus, the first one is used to study the response of long samples with arbitrary cross section to transverse magnetic fields[3] [2, 3] and also in levitation systems with translational symmetry [4]. The second one is used to describe magnetic properties in several realistic configurations such as magnetisation measurements with fields along the axis of finite cylinders or magnetic levitation experiments with aligned cylindrical magnets and superconductors.

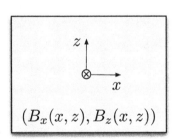

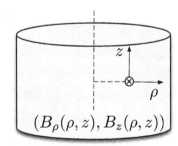

Figure 8.1. Schematics of the 2D problems considered in this chapter. To the left, we show the cross section of a long rod with magnetic field in the transverse plane. To the right we display a cylinder in a rotationally symmetric problem.

---

[3] For example in the so-called 'flux shaking experiments' [2].

8.1.2 Discrete formulation

In the forthcoming sections, we will put forward several examples within the class of problems sketched in figure 8.1. As numerical procedures will be introduced, we will first give some details about the discrete form of the above equations. The basic concept for discretisation is that, having disregarded the charge accumulation processes (recall the magnetoquasistatics (MQS) condition $\nabla \cdot \mathbf{J} = 0$), the currents flowing within the sample may be treated as a collection of n circuits, each with well defined current, say $J_1, J_2, \dots, J_n$. In a general framework, identifying such circuits *a priori* may be a non simple task. However, aided by the above mentioned symmetry considerations, in the context of this chapter this will not be a difficulty, the shape of circuits will be known[4]. For the interested reader, in chapter 10 we will concentrate on the problem of extending these ideas to problems for which the circuits themselves are additional unknowns.

As introduced in section 5.2.2 a rather convenient representation for our problem will be to 'gather' the values of the circuital currents in a multidimensional vector that will be expressed in the form

$$|\mathbf{J}\rangle \equiv \begin{pmatrix} J_1 \\ J_2 \\ \cdot \\ \cdot \\ J_n \end{pmatrix}, \qquad \langle\mathbf{J}| \equiv (J_1, J_2, \dots, J_n) \tag{8.3}$$

The components J_i of the *grid vector* $|\mathbf{J}\rangle$ will be nothing but the unknowns of the problem. Obviously, the higher the value of n, the higher the resolution, but also the computational cost. Recall that these vectors are not to be confused with the 3-dimensional space vector components of the physical quantities.

Tightly related to the concept of discretization is the definition of inductance. Recalling the celebrated Neumann's formula for the mutual inductance between two circuits

$$M_{ij} = \frac{\mu_0}{4\pi} \oint_{c_i} \oint_{c_j} \frac{d\ell_i \cdot d\ell_i}{\|\mathbf{r}_i - \mathbf{r}_j\|} \tag{8.4}$$

one can show that, for instance, $\mathcal{F}_{JJ}$ may be written as

$$\mathcal{F}_{JJ} = \frac{1}{2}\langle\mathbf{J}|\mathbf{m}|\mathbf{J}\rangle \tag{8.5}$$

Here, we introduce the $n \times n$ inductance matrix $\mathbf{m}$. As for the case of the vector of 'circuital' currents, lower case notation is used in order to emphasise that normalised (dimensionless) quantities are involved. Normalisation factors, as well as the specific expressions to be used for the evaluation of the elements m_{ij} will be defined for each particular case and computer code presented in the forthcoming sections.

---

[4] For the translational symmetry, the circuits may be thought of as parallel long filaments, and for the rotational symmetry will be circular loops.

Analogously, one may obtain the expressions for $\mathcal{F}_{J0}$ and $\mathcal{F}_{JS}$

$$\mathcal{F}_{J0} = \langle \mathbf{J}^{\vee} | \mathbf{m} | \mathbf{J} \rangle$$
$$\mathcal{F}_{JS} = \langle \Delta \psi_S | \mathbf{J} \rangle \tag{8.6}$$

For compactness, we introduce the notation $|\mathbf{J}^{\vee}\rangle$ for the vector with (already known) values of current at the previous time layer. $|\psi_S\rangle$ contains the vector potential due to the external sources at the localisation of each circuit, i.e. ψ_{S_i} will be related to the vector potential acting on the circuit of current J_i and occasionally including some metric factor. For the symmetries of interest, the vector potential will be constant along each circuit, and parallel to the flow of current. Finally, for each case, the term $\Delta t \, \mathcal{W}_{JE}$ will be specified separately.

In conclusion, the evolution of the current density $\mathbf{J}(\mathbf{r}, t)$ within the superconductors under a given excitation process specified by $\mathbf{A}_S(\mathbf{r}, t)$ will be obtained by the discretised version of equation (8.1) along such process, i.e.

$$\texttt{minimise} \left[\frac{1}{2} \langle \mathbf{J} | \mathbf{m} | \mathbf{J} \rangle - \langle \mathbf{J}^{\vee} | \mathbf{m} | \mathbf{J} \rangle + \langle \Delta \psi_S | \mathbf{J} \rangle + \Delta t \, \mathcal{W}_{JE} \right]$$

As shown below, this statement may be easily solved with the help of MATLAB and OCTAVE dedicated software that implements flexible built-in functions for this kind of problem (see chapter 6.1 for a basic introduction to the specific tools). Throughout the forthcoming sections, the reader will be guided through several examples that provide interactive solutions in various configurations. For all cases, the workflow will consist of:

1. Define the geometry of the problem and identify the actual circuits for the flow of current.
2. Obtain the expressions for the related induction matrix elements m_{ij}.
3. Define the dissipation process of interest, i.e. the dependence $\mathcal{W}_{JE}(|\mathbf{J}\rangle)$.
4. Define a certain 'experimental' process, typically by the dependence $\mathbf{A}_S(\mathbf{r}, t)$.
5. Use the minimisation procedure in order to update the current vector step by step.
6. Analyse data.

8.2 The critical state model: transport problem

To start with, we put forward an example that falls rather into the academic side. It will be used to gain familiarity with the tools of analysis, and later to illustrate some 'special' situations concerning the actual dissipation process (i.e. the dependence $\mathcal{P}(\mathbf{J})$) and the definition of the external sources.

Let us suppose that a transport current is driven along the y-axis of a superconducting slab as sketched in figure 8.2. The slab is defined by

$$|x| < \infty, \quad |y| < \infty, \quad |z| < d \tag{8.7}$$

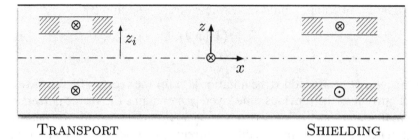

Figure 8.2. Sketch of the superconducting slab (infinite along the x and y-axes) To the left, a generic layer and its symmetric are shown for the configuration of transport current along y-axis. To the right, the situation for a configuration with applied magnetic field along x-axis.

Assuming that the slab behaves according to Bean's critical state model, the reader can straightforwardly obtain the CS solution (recall section 3.3). When a transport current of value K per unit length (across x) flows along the y-axis[5], it penetrates uniformly from both sides of the slab to a depth given by

$$z_0 = d - \frac{K}{2J_c} \tag{8.8}$$

with J_c being the critical current density of the bulk.

Apparently, a maximum current of value $K_c = 2dJ_c$ is allowed in the superconducting state. For $K < K_c$ one can easily construct the evolutionary penetration profiles of an AC current by combining surface layers with current density $\pm J_c$. Taking advantage of this simplicity we will introduce (and test) the basic protocol and numerical tools for further use in complex problems. We state the numerical problem as follows.

8.2.1 Geometry and circuits

As sketched in figure 8.2, for the discretisation of the problem, the elements of interest will be n layers of width d/n, each centred at the position z_i and carrying a maximum current (critical state model ansatz) per unit length, i.e.

$$|k_i| \leqslant \frac{J_c d}{n} \tag{8.9}$$

In dimensionless units $J_i \equiv n k_i / J_c d$ one has

$$|J_i| \leqslant 1 \tag{8.10}$$

Incidentally, we note that, owing to the symmetry of the problem, we are only considering the upper part ($z > 0$), because the same pattern will occur for $z < 0$.

---

[5] This is nothing but the so-called 'sheet current' that in a more general situation may be calculated as $K(x) = \int J_y(x, z) \, dz$.

8.2.2 Inductance matrix

In this case of infinite geometry, the evaluation of the matrix elements m_{ij} is better done by straightforwardly evaluating the magnetostatic energy from the expression of its density $B^2/2\mu_0$ and relating B to the current density by applying Ampère's law. Thus, considering that B is the superposition of layer by layer contributions, the internal[6] energy per unit area of slab may be written as

$$\frac{1}{2\mu_0} \int B^2 dz = \frac{\mu_0}{2}\left(\frac{d}{n}\right)^3 J_c^2 \sum_{i=1}^{n}\left[\frac{J_i}{2} + \sum_{j<i} J_j\right]^2 \qquad (8.11)$$

Based on this, and up to the dimensional prefactor that may be absorbed in the expression of $\mathcal{F}$, one has[7]

$$m_{ij} = \begin{cases} 2(1/4 + n - i), & i = j \\ 1 + 2(n - \max\{i, j\}), & i \neq j \end{cases} \qquad (8.12)$$

8.2.3 Dissipation mechanism

We recall that, in the form exposed by Bean (section 3.3), the critical state model may be interpreted in terms of an arbitrary large slope in the $E(J)$ curve as soon as J goes beyond J_c, and this gives way to a regime of strong dissipation. As a result of this, the excursions of J to higher values, instantaneously revert back to J_c, while penetration within the undercritical areas $(0 \to J_c)$ occurs. Mathematically, one can implement this physical behaviour by replacing the term $\Delta t\, \mathcal{W}_{JE}$ with the inequality condition

$$J \leqslant J_c \Rightarrow |j_i| \leqslant 1 \quad \forall i \qquad (8.13)$$

that will be applied when minimisation is performed. This is easily (in fact, automatically) implemented by means of the proposed software.

For the readers' sake, later in this chapter, we will explicitly show that replacing $\Delta t\, \mathcal{W}_{JE}$ by the inequality constraint may be understood as a limiting situation of a continuous process that assumes steeper $E(J)$ laws.

8.2.4 Experimental process

Another feature of the current example is the kind of external constraint used. Thus, a total 'transport current' gives way to a restriction of the form[8]

---

[6] The energy outside the slab is obviously diverging owing to the unphysical situation for an infinite sample. However, this mathematical drawback is not a problem. The outer energy is a constant however the circulating current distributes within the slab, and so it is irrelevant for the variational problem.

[7] Start by squaring the bracketed expression, observe the first elements and group terms in order to recognise the coefficients for each combination of j_i and j_j.

[8] The factor 2 comes from the fact that, by symmetry one has the symmetric profile $k(z_i) = k(-z_i) \equiv k_i$.

$$2 \sum_{i=1}^{n} k_i = K \tag{8.14}$$

which in dimensionless units reads

$$\sum_{i} J_i = \frac{K}{2J_c d} \, n \tag{8.15}$$

This will replace the term $\mathcal{F}_{\mathrm{JS}}$ in the general statement.

8.2.5 Minimisation

As a conclusion of the above arguments, the evolutionary minimisation statement for the critical state model with transport current is

$$\begin{cases} \texttt{minimise} \left[\frac{1}{2} \langle \mathbf{J} | \mathbf{m} | \mathbf{J} \rangle - \langle \mathbf{J}^{\vee} | \mathbf{m} | \mathbf{J} \rangle \right] \\ \texttt{for } |J_i| \leqslant 1 \\ \texttt{and } \sum_{i=1}^{n} J_i = \frac{K}{2J_c d} \, n \end{cases} \tag{8.16}$$

This formulation gives the step-by-step evolution of the system subjected to some external excitation process. i.e.

$$K(t_n) \to K(t_n + \Delta t) \Rightarrow |\mathbf{J}^{\vee}\rangle \to |\mathbf{J}\rangle \tag{8.17}$$

In general, one has to be careful about the smallness of Δt in order to ensure a meaningful application of the variational theory (recall section 1.2.4). It is only for a few exceptions that this condition may be overlooked.

8.2.6 Application and analysis

Let us apply the minimisation principle to the following process for the applied transport current

$$K: 0 \to 1.6 J_c d \to -1.6 J_c d \tag{8.18}$$

Notice that this means that the transport current is first ramped up to 80% of the maximum value allowed by the superconductor without dissipation, and then ramped down to the same percentage but flowing in the opposite direction. The results that may be obtained with the help of the accompanying software are displayed in figure 8.3.

There, we plot the solution $J_i (1 = 1, \ldots, n)$ for different instants along the process. As expected, when the applied current is incremented, a layer of current density J_c penetrates deeper, while the associated magnetic field displays constant slope. If the applied current reverts values, a surface layer of density $-J_c$ advances towards the sample centre and the magnetic field displays opposite slope. Figure 8.3 has been created with the help of the following codes

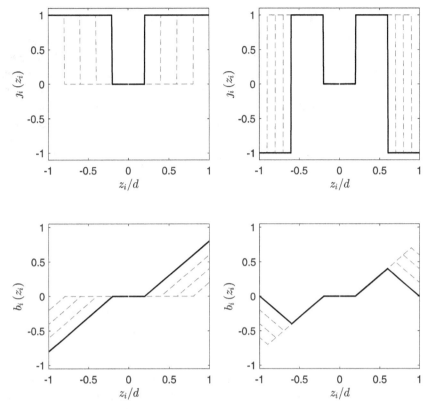

Figure 8.3. Dimensionless current density and magnetic field profiles (upper and lower panels respectively) within the superconducting slab (figure 8.2) when a transport current is cycled along the y-axis. Units are established by J_c and $\mu_0\,J_c\,d$. Thick lines correspond to the profiles for which the current reaches a maximum amplitude ($K_m = 1.6 J_c\,d$) and reverts to zero ($K = 0$). Dashed lines are for intermediate steps ⌀. Source codes available at https://iopscience.iop.org/book/978-0-7503-2711-4.

- `slab_cs_qp_transport.m` (main)
- `matrixMXt.m`; `selfXt.m`; `mutualXt.m`;
- `plot_profile.m`; `subplot_current.m`; `subplot_field.m`

The main program defines the process and performs the sequential minimisation steps. Taking advantage of the fact that the problem is quadratic in the unknowns, it makes use of the very efficient functions `quadprog` (MATLAB) and `qp` (OCTAVE). For convenience, results are stored in data files that may be later analysed or used to restart calculations. The matrix elements m_{ij} are evaluated in independent scripts that are called from the main program. Data analysis (including the calculation of the magnetic field based on equation (8.11)) and desired plots may be done *a posteriori* upon loading the required data files (one per time step in the process).

8.3 The critical state problem: magnetisation

In this case we concentrate on the response of a superconducting sample in the critical state to an external applied magnetic field. The process will be encoded through some dependence $A_S(t)$ and introduced in the minimisation principle. On the other side, we will use a finite geometry, quite close to the experimental situations. The sample will be a circular cylinder of arbitrary aspect ratio subject to a uniform field along its symmetry axis. In this case, a simple analytical theory is not available. Nevertheless, with the tools provided below, the reader may explore the whole range and, in particular, check the analytical expressions available in limiting situations, i.e. discs and long cylinders. Noticeably, the variational statement is applied indistinctly to any arbitrary value of the aspect ratio.

8.3.1 Geometry and circuits

As shown in figure 8.1, we consider a cylindrical sample that will be subject to an applied magnetic field that is uniform and applied along its axis. Apparently, the induced currents will flow in the form of circular loops. This is sketched in figure 8.4 that outlines the localisation of the ith circuit, i.e. a loop of radius ρ_i at the height z_i. Thus, our numerical task will consist of defining a set of virtual loops, each carrying an unknown current that will have to be determined through the solution of some set of equations.

Below, we put forward a classification scheme that allows to deal with the 2D geometry with relative ease. Assume that figure 8.4 represents the section of a cylindrical sample of radius R and height L. If one uses a square grid of points with interval δ, the full problem will be stated in terms of $N \times M$ circuits with $N = R/\delta$ and $M = L/\delta$ (safe small rounding factors that will be less important as δ becomes smaller). Then, if one sets the origin of coordinates at the centre of the basis the cylinder, the full grid of points may be classified by a single index. In normalised units

$$\tilde{\rho}_i \equiv \frac{\rho_i}{\delta} = i + N\left(1 - \lceil \frac{i}{N} \rceil\right)$$

$$\tilde{z}_i \equiv \frac{z_i}{\delta} = \lceil \frac{i}{N} \rceil$$

(8.19)

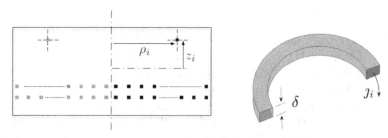

Figure 8.4. Sketch of the cylindrical superconductor in the axisymmetric problem. The induced circular loops of current are sketched. (ρ_i, z_i) indicates the position of the generic ith circuit (loop of radius ρ_i at position z_i). To the right we show the current j_i circulating along the generic loop with square cross section of size $\delta \times \delta$ and centred at (ρ_i, z_i).

Here, we have used the standard notation for the ceiling functions, i.e. $\lceil z \rceil = \min\{k \in \mathbb{Z} \mid k \geqslant z\}$, and $i = 1, 2, \ldots, n \equiv N \times M$. For convenience, one could eventually symmetrise by subtracting the half height: $\tilde{z}_i \to \tilde{z}_i - L/2\delta$.

8.3.2 Inductance matrix

The mutual inductance between two circular loops is a classical problem that has been dealt with in many texts. In fact, the formulas below were already derived by Maxwell himself. A rather famous reference with many useful expressions is the book by Grover [5][9]. Thus, the mutual inductance coefficient between two coils of negligible cross section is given by

$$M_{ij}^{\circ} = \mu_0 \sqrt{\rho_i \rho_j} \left[\left(\frac{2}{k} - k \right) K(k) - \frac{2}{k} E(k) \right] \tag{8.20}$$

with the definition

$$k = \sqrt{\frac{4\rho_i \rho_j}{(\rho_i + \rho_j)^2 + (z_i - z_j)^2}}$$

and K, E the complete elliptic integrals of the first and second kind. The self inductance of the loops is given by

$$M_{ii}^{\circ} = \mu_0 \rho_i \left[\ln \frac{\rho_i}{\delta} + 0.9018 \right] \tag{8.21}$$

8.3.3 Dissipation mechanism

As in the previous example we will use the inequality constraint for the current density, i.e. $J_i \leqslant 1$. In this case, the dimensionless units for the current density are given by

$$J_i \equiv \frac{J(\rho_i, z_i)}{J_c} \tag{8.22}$$

i.e. J_i is the normalised current that flows along the circular loop with cross section δ^2 centred at the point (ρ_i, z_i).

8.3.4 Experimental process

An applied uniform magnetic field along the axis of the cylinder may be represented by the vector potential

---

[9] Since the first edition in 1946, this book has been a classical reference in the field. A lot of practical expressions may be found for the self and mutual inductance coefficients of many geometries. It may be very useful for the numerical implementation of the minimisation principle, that relies on the inductance matrix formalism.

$$\mathbf{A}_S = \frac{\mu_0}{2} H_a \, \rho \, \hat{\boldsymbol{\phi}} \equiv A_S(\rho) \, \hat{\boldsymbol{\phi}} \tag{8.23}$$

Then, the physical process under consideration will be specified by some dependence $H_0(t)$, for instance a typical case of interest is the hysteresis loop type:

$$H_a: 0 \rightarrow H_m \rightarrow -H_m \rightarrow H_m \tag{8.24}$$

for which the applied field is ramped up and down with an amplitude H_m.

8.3.5 Minimisation

By using dimensionless units defined as follows

$$m_{ij} \equiv M_{ij}(\tilde{\rho}_i, \tilde{\rho}_j; \tilde{z}_i, \tilde{z}_j)/\mu_0$$
$$\psi_i^{S} \equiv \pi \tilde{\rho}_i^{2} h_0 \tag{8.25}$$
$$h_a \equiv H_a/(J_c \, \delta)$$

one ends up with the following variational principle[10]

$$\begin{cases} \text{minimise} \left[\frac{1}{2}\langle \mathbf{J}|\mathbf{m}|\mathbf{J}\rangle - \langle \mathbf{J}^{\vee}|\mathbf{m}|\mathbf{J}\rangle + \langle \Delta \boldsymbol{\psi}^{S}|\mathbf{J}\rangle \right] \\ \text{for } |J_i| \leqslant 1 \qquad \forall i \end{cases} \tag{8.26}$$

Again, we recall that this formulation gives the step-by-step evolution of the system as the response to the external excitation process, i.e.

$$H_a(t_n) \rightarrow H_a(t_n + \Delta t) \Rightarrow |\mathbf{J}^{\vee}\rangle \rightarrow |\mathbf{J}\rangle \tag{8.27}$$

There are essential properties that may be straightforwardly studied with the accompanying software. They relate to the response of cylindrical samples with arbitrary aspect ratio, a topic that has been of much interest in the scientific literature [6, 7]. Noteworthily, this geometry relates to an enormous set of experiments with samples shaped as cylinders, but ranging from discs to (ideally infinite) long cylinders. Figure 8.5 displays the main feature of flux penetration in the finite cylinder geometry. Magnetic flux and the accompanying critical currents penetrate defining a curved front that separates the critical region and the flux free core. By changing the parameters of the simulation, the reader may check that the front will become flatter as the sample tends to the shape of a long cylinder. Concerning the magnetic flux lines, we recall that, in 2D problems the lines of magnetic field coincide with the contour plot of the vector potential. This property is used in our code, that eventually calls the MATLAB function contour() and displays isolines of the full vector potential $\mathbf{A} = \mathbf{A}_S + \mathbf{A}_{SC}$. As proposed in exercise R8-3, the contribution of the superconductor $\mathbf{A}_{SC}$ may be straightforwardly evaluated in terms of the mutual inductance matrix conveniently redefined.

---

[10] Here, a global factor with units of energy $\mu_0 J_c^2 \, \delta^5$ has been overridden in the minimisation. Also, the function $\psi = \rho A_S(\rho)$ has been introduced for convenience in cylindrical geometry.

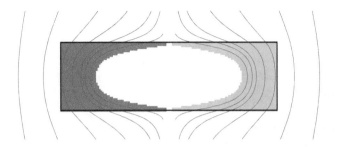

Figure 8.5. Magnetic field lines and penetration profile for a cylindrical superconductor ($2R/L = 3.3$) under uniform magnetic field along its axis in the partial penetration regime. The shaded area corresponds to $J = J_c$, in contrast with the flux free regime ($J = 0$). Different colours in the plot indicate 'inwards' or 'outwards' current. Source codes available at https://iopscience.iop.org/book/978-0-7503-2711-4.

8.3.6 Application and analysis

Below, we analyse the magnetic response of two different cylindrical superconductors to an axial uniform field that is cycled according to equation (8.24). For each geometry, we choose three values of H_m, viz.: $H_m = 0.5H_p$, H_p, $2H_p$ where $H_p \equiv J_c R$. Recall (chapter 3.3) that H_p is the full penetration field, a quantity that characterises the saturation of the ideally infinite (very long) cylindrical superconductor. The results are displayed in figure 8.6, where we plot the normalised volume magnetisation, again established by the saturation value, i.e. $M_p = J_c R/3$, that corresponds to the cylinder fully penetrated by circular circuits of superconducting current with maximal density J_c. Here, magnetisation is evaluated in terms of the overall magnetic moment $\boldsymbol{\mu}$:

$$M_v = \mu/\pi R^2 L = \left[\frac{1}{2}\int_\Omega (\mathbf{r} \times \mathbf{J})\, dV\right]/\pi R^2 L \qquad (8.28)$$

Numerically, in order to perform this operation, we will sum over circular circuits of normalised radii $\tilde{\rho}_i$, each carrying a normalised current J_i

$$\frac{M_v}{M_p} = 3\frac{\delta^5}{R^3 L}\sum_{i=1}^{N\cdot M}\tilde{\rho}_i^2\, J_i = \frac{1.5}{N^3 M}\sum_{i=1}^{N\cdot M}\tilde{\rho}_i^2\, J_i \qquad (8.29)$$

Figures 8.5 and 8.6 have been created with the help of the following codes
- `cyl_cs_qp.m` (main)
- `generate_grid.m`
- `matrixM.m`; `self.m`; `mutual.m`; `psi_pot_unif.m`
- `profile_Jcyl_lines.m`; `matrixM_augmented.m` (figure 8.5)
- `magnetisation_loops.m`; `e_cylinder.m` (figure 8.6)

The main program defines the problem and performs the sequential minimisation process. Again, taking advantage of the fact that the problem is quadratic in the

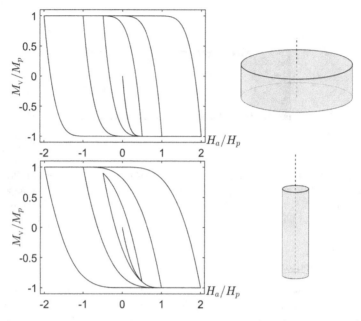

Figure 8.6. Magnetisation loops corresponding to cylindrical samples of different aspect ratios ($2R/L = 3.33$ and $2R/L = 0.3$ respectively) as shown in the plot. For each case, three values of the maximum applied field have been used, i.e. $H_m = 0.5H_p$, H_p, $2H_p$. Dimensionless units are defined in the text $\mathscr{O}$. Source codes available at https://iopscience.iop.org/book/978-0-7503-2711-4.

unknowns, we use of the functions `quadprog` (MATLAB) and `qp` (OCTAVE). For each time step, results are stored in a data file that may be later analysed or used to restart calculations. The coordinates of points within the cylinder are generated for the given combination of radius and length (function `generate_grid`), and stored in one dimensional arrays for further processing. The inductance matrix elements m_{ij} are evaluated in independent scripts that are called from the main program. Eventual data analysis (including the calculation of the magnetic field lines, evaluation of flux penetration profiles and sample magnetisation) may be done *a posteriori* upon loading the required data files and utilising the codes of the last two groups.

The results clearly show the expected behaviour. On the one hand, both sample geometries reach the exact saturation condition $M_v = M_p$ (each for a certain value of the applied magnetic field). Recall that M_p is a reference value independent of the cylinder's aspect ratio, because it just 'measures' the magnetic moment per unit volume in a saturated sample. On the other hand, as one approaches the long sample limit $L \gg R$, saturation occurs closer to the field $H_a = H_p$, whereas for $L < R$ magnetisation saturates for lower and lower fields. Thus, in the upper panel of figure 8.6, one has $M_v = M_p$ at the field $H_a = 0.5H_p$. The reader can check that when the sample approaches the 2D limit (disk geometry, $L \ll R$) the hysteresis cycle becomes a rectangle.

8.4 Response to non-uniform magnetic fields

As in the previous section, here we will study the response of a cylindrical superconductor, in the critical state, to some external process described by a certain dependence $A_S(\mathbf{r}, t)$. In this case, the source will be a neighbouring permanent magnet of cylindrical shape with its axis coincident with the superconductor's. Thus, although non-uniform, the distribution of fields will still be cylindrically symmetric, and the current circulation will still be along circular paths (figure 8.4). Indeed, one may follow exactly the same protocol described in section 8.3 with two unessential exceptions: (i) $A_S(\mathbf{r}, t)$ will be expressed in terms of the permanent magnet's properties, and (ii) the process of interest will be described in terms of some *route* of the permanent magnet in the neighbourhood of the superconductor, i.e. cooling at some distance, increasing or decreasing separation, etc. In fact, this configuration corresponds to a great number of experiments testing the levitation properties in superconductor–magnet devices, as will be further discussed in chapter 13. Below, we will concentrate on the incorporation of these two aspects.

8.4.1 The permanent magnet's vector potential

Let the permanent magnet be a cylinder of radius R_m and length L_m, uniformly magnetised along its axis, i.e.

$$\mathbf{M} = M_0 \hat{k} \tag{8.30}$$

The corresponding vector potential admits a simple representation if one uses a well known physical picture. The magnet is *equivalent* to a surface current distribution with density

$$\mathbf{K}_m = \mathbf{M} \times \hat{n}, \tag{8.31}$$

$\hat{n}$ being the unit vector normal to the magnet's surface at each point. Then, in our discretised picture, the magnet may be replaced by an equivalent solenoid, formed by N_m parallel circular coils of radius R_m (strictly speaking $R_m - \delta/2$) extended along the length L_m. Each coil will have a square section $\delta \times \delta$,[11] so that, the number of required coils will be

$$N_m = \frac{L_m}{\delta} \tag{8.32}$$

Eventually, the vector potential acting on a certain superconducting circuit may be calculated in terms of mutual inductances. More specifically, the related function ψ (by analogy to equations (8.25) and (8.26)) is given by

$$\psi_i^{Sm} = \frac{M_0}{J_c \, \delta} \sum_{\alpha=1}^{N_m} \mathbf{m}_{\alpha i} \tag{8.33}$$

---

[11] In fact, the actual section of the *equivalent coils* could be optimised for each case, depending on the aspect ratio of the magnet. Here for simplicity, we choose the same square grid that is used for the superconductor.

with $\mathbf{m}_{\alpha i}$ being the elements of the normalised mutual inductance matrix between the ith superconducting loop and the αth loop in the equivalent magnetic solenoid.

From the physical point of view, it is interesting to notice that the relevance of the interaction between the magnet and the superconductor is gauged by a dimensionless parameter, that compares values of magnetisation (in physical units $p \equiv M_0/J_c R$). In conclusion, the penetration of magnetic flux within the superconductor will be obtained through the principle

$$\begin{cases} \text{minimise} \left[\frac{1}{2}\langle\mathbf{J}|\mathbf{m}|\mathbf{J}\rangle - \langle\mathbf{J}^\vee|\mathbf{m}|\mathbf{J}\rangle + \langle\Delta\boldsymbol{\psi}^{Sm}|\mathbf{J}\rangle\right] \\ \text{for } |j_i| \leqslant 1 \quad \forall i \end{cases} \tag{8.34}$$

8.4.2 Application and analysis

As deduced from figure 8.7, we will provide some tools that allow to analyse a process in which the superconductor is cooled down past the transition temperature in the presence of a bias magnetic field produced by a permanent magnet at some initial distance z_0. Subsequently, the magnet undergoes a displacement along the symmetry axis. The plot has been created under realistic conditions as concerns the relevant physical parameters in many experiments: (i) for the superconductor, we have taken $R = 25$ mm and $L = 15$ mm, (ii) for the magnet $R_m = 22.5$ mm and $L_m = 15$ mm, (iii) we have assumed the value $J_c = 3 \times 10^8$ A/m^2 for the critical current density, a reasonable expectation for a melt-textured YBaCO bulk at $77K$, and (iv) corresponding to a standard NdFeB magnet, we have used $\mu_0 M_0 = 1.17$ T. On the other hand, the process leading to this plot starts by cooling the superconductor with the magnet at a distance $z_0 = 2R$.

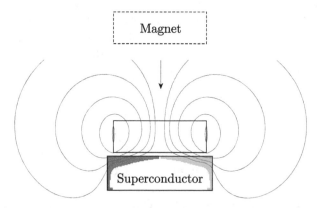

Figure 8.7. Magnetic field lines around a magnet/superconductor cylindrical arrangement. As indicated, the magnet was held at a long distance in the cooling process and subsequently displaced towards the superconductor. The region penetrated by screening currents is marked ($J = J_c$). Colours indicate inward/outward current in the cross section of the plot. Realistic physical parameter values have been used (see text) $\mathcal{O}$. Source codes available at https://iopscience.iop.org/book/978-0-7503-2711-4.

Experimentalists would customarily consider that, in the above conditions one may assume a ZFC process[12] and neglect any trapped field in the superconductor. As the reader may check, this is confirmed by the simulations. The weak magnetic field associated with the magnet in the initial position barely generates induced superconducting currents when the magnet initiates the movement. In brief, this means that a cooling distance of any higher value for z_0 would make no difference, which allows to declare ZFC conditions. The reader may check that, as z_0 becomes smaller and smaller (FC conditions), frozen magnetic field profiles within the superconductor are more and more influential. This topic will be thoroughly covered in chapter 13. Figure 8.7 has been created with the help of the following codes

- `cyl_cs_lev_qp.m` (main)
- `generate_grid.m`
- `matrixM.m`; `self.m`; `mutual.m`; `psi_pot_magnet.m`
- `profile_Jcyl_lines_magnet.m`; `matrixM_augmented_magnet.m`

The basic difference to their counterparts in section 8.3 is the introduction of the magnet's vector potential, and the definition of the actual process. In this case, the 'reaction' of the superconducting cylinder arises from the displacement of the nearby magnet.

8.5 Finite resistivity: *piece-wise* approximation

In this section, we start dealing with the topic of introducing a finite dissipation law in the critical state related variational problem. Physically, the inclusion of such effect in the vortex state of superconductors may be required by the experimental conditions of interest. However, it is noteworthy that, depending on the material and on the experimental range, different models will have to be used. In this section, we will provide the model that applies to the *flux flow* regime. As said, this occurs at low temperatures with negligible thermal creep of flux lines, and when the threshold value of the critical current density is noticeably overcome. In such case, dissipation may be successfully interpreted in terms of the so-called flux flow resistivity ρ_f (see equation (2.47)) that will be a new parameter in our models.

To start with, we recall the transport problem in the slab geometry in section 8.2. Here, we also consider the situation described in equation (8.17), but include a dissipation term in the process.

We apply the 'minimal' modification of the critical state, that uses a *piece-wise* conduction law[13]

---

[12] ZFC (zero field cooling conditions) and FC (field cooling) are very relevant in high pinning materials. In the latter case, when the sample is cooled down, vortices nucleate and pin according to the applied field structure.
[13] The actual law used in this section applies to 1D problems (slab and infinite cylinder with transport current). For those situations in which the local orientation of the vector **J** is not known *a priori* further considerations are needed, as will be dealt with in chapter 10.

$$E(J) = \begin{cases} \rho_f\,(J + J_c), & J < -J_c \\ 0, & -J_c \leqslant J \leqslant J_c \\ \rho_f\,(J - J_c), & J_c < J \end{cases} \qquad (8.35)$$

From the physical point of view, this approximation means that vortices remain static until a certain force threshold is reached (for $J = J_c$). Beyond this point, the driving forces overcome the pinning reaction, flux lines drift and flux flow resistivity appears.

Recalling the notation introduced in section 1.2, on may obtain the quasi-steady transport regime by introducing the dissipation function $\mathcal{P}$ through the relation

$$\mathbf{E} = \nabla_{\mathbf{J}}\mathcal{P} \Rightarrow \mathcal{P} = \int \mathbf{E}(\mathbf{J}) \cdot d\mathbf{J} \qquad (8.36)$$

In the case under consideration, this leads to

$$\mathcal{P}(J) = \begin{cases} \rho_f(J + J_c)^2/2 & , \quad J < -J_c \\ 0 & , \quad -J_c \leqslant J \leqslant J_c \\ \rho_f(J - J_c)^2/2 & , \quad J_c < J \end{cases} \qquad (8.37)$$

This piece-wise definition may be dealt with in several forms. For instance, one can make use of the step function $\Theta(J)$. A useful representation is based on the relation

$$\Theta_c(J) = \lim_{k \to \infty} \frac{1 + \tanh[k((J/J_c)^2 - 1)]}{2} \qquad (8.38)$$

As the reader may check, in practice, taking $k \gtrsim 20$ performs rather well.

Thus, as sketched in figure 8.8, the required dissipation function (according to equation (8.37)) is simply obtained multiplying by Θ_c, i.e.

$$\mathcal{P}(J) = \frac{\rho_f}{2}\Theta_c(J)(J \mp J_c)^2 \qquad (8.39)$$

with '$\mp$' applying respectively to $J > 0$ and $J < 0$.

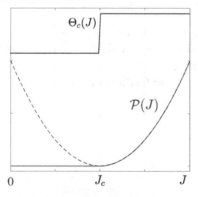

Figure 8.8. Representation of the dissipation function $\mathcal{P}(J)$ obtained by multiplying the parabolic relation $\rho_f(J - J_c)^2/2$ by a step function $\Theta_c(J)$ as shown in the plot.

8.5.1 Minimisation

The discretised minimisation principle will include a term of the form

$$\Delta t \int_{\Omega} \mathcal{P}(J)dV \rightarrow \frac{1}{2}\Delta t \sum_{i=1}^{n} r_i\, \Theta_c(J_i)(J_i \mp 1)^2 \qquad (8.40)$$

written in terms of the normalised current J_i and the resistance r_i for each circuit of the problem. In our case, r will be a constant. Nevertheless, one may incorporate inhomogeneous behaviour easily, just by defining an array of values r_i.

After some algebra, our material law takes the form

$$\begin{cases} \texttt{minimise} \left[\dfrac{1}{2}\langle\mathbf{J}|\mathbf{m}|\mathbf{J}\rangle - \langle\mathbf{J}^{\vee}|\mathbf{m}|\mathbf{J}\rangle + \langle\mathbf{J}|\mathbf{P}|\mathbf{J}\rangle \mp 2\langle\mathbf{1}|\mathbf{P}|\mathbf{J}\rangle\right] \\ \texttt{for } \displaystyle\sum_{i=1}^{n} J_i = \dfrac{K}{2J_c d}\, n \end{cases} \qquad (8.41)$$

Here, we have introduced the notation $\langle\mathbf{1}|\equiv(1, 1, \ldots , 1)$ and defined the dissipation matrix $\mathbf{P}$ with elements

$$p_{ij} = \delta_{ij}\,\frac{\Delta t\,\rho_f}{\mu_0 d^2}\,\Theta_c(J_i) \qquad (8.42)$$

8.5.2 Application and analysis

We call the readers' attention that, differently to the previous cases in this chapter, the above minimisation principle (8.41) does not restrict the admissible values of the individual unknowns, i.e. the condition $J_i \leqslant 1$ has been released. Physically, we are relaxing the limiting (though frequently appropriate) condition of infinite dissipation beyond J_c. This opens the possibility of describing new phenomena. In particular, having included the variable Δt in the formulation, we have the chance to investigate time relaxation effects.

Figure 8.9 shows an example of how one can apply the principle (8.41) for investigating the evolution of the transport current penetrating within the super-conducting slab (figure 8.2) with an $E(J)$ law given by equation (8.35). The simulation proceeds as follows. One sets a 'final' value of the transport current (K) and iteratively obtains the current density profiles (values of J_i) by successive minimisation steps using equation (8.41), each corresponding to a small time increment Δt:

$$\begin{aligned} &\texttt{Set } \sum_i J_i \quad (\texttt{Value of } K) \\ &\downarrow \\ &\Delta t \Rightarrow \quad J_1^{(1)}, J_2^{(1)}, \ldots , J_n^{(1)} \\ &\downarrow \\ &2\Delta t \Rightarrow \quad J_1^{(2)}, J_2^{(2)}, \ldots , J_n^{(2)} \\ &\vdots \\ &N\Delta t \Rightarrow \quad J_1^{(N)}, J_2^{(N)}, \ldots , J_n^{(N)} \end{aligned} \qquad (8.43)$$

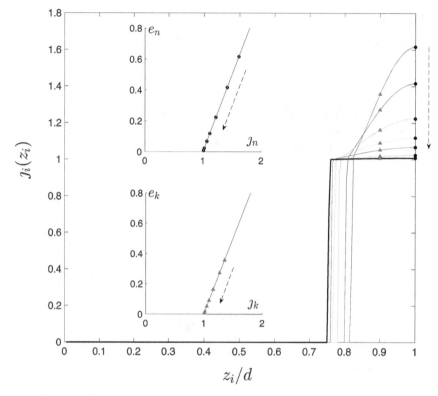

Figure 8.9. Time relaxation of the current density penetration profile within the superconducting slab after applying a transport current step. Only the upper half is depicted ($z_i \geqslant 0$). The insets show the electric fields evaluated by means of equation (8.35) at two points: z_k within the sample and z_n at the very surface 🔗. Source codes available at https://iopscience.iop.org/book/978-0-7503-2711-4.

When this process is finished, one updates K, performs the relaxation steps again, and so on.

As one can observe in the plot, an overcritical region $J_i > 1$ is generated close to the surface of the sample. Then, progressively, the current density 'relaxes' down towards the critical value and, eventually, the equilibrium state corresponding to the specific transport value K is established. This is characterised by the achievement of the condition $e = 0 \Leftrightarrow J = 1$, i.e. the transient electric field goes to zero when the current density reaches the critical value, and no further evolution occurs if the external source condition does not change.

Apparently, the establishment of the eventual critical state profile for a given increment of transport current will be possible for a certain relation of time constants. In the above numerical simulation, it is apparent that allowing a smaller number of intermediate steps would imply an incomplete formation of the critical state profile. Considering a realistic situation, a basically complete relaxation will be possible if the following condition is satisfied

$$N\Delta t \gg \frac{\mu_0 d^2}{\rho_f} \tag{8.44}$$

which means that the time constant related to the external source variation is much above the characteristic *diffusion* time constant, given by the sample's dimensions and resistivity. Figure 8.9 has been created with the help of the following codes

- `slab_PW_qp_transport.m` (main)
- `matrixMXt.m`; `selfXt.m`; `mutualXt.m`
- `relax_PW_profile.m`; `createfigure_PW.m`

that follow the same procedures defined in their counterpart of section 8.2 with two distinctive features: (i) first, the new minimisation algorithm does not include the restriction $J_i \leqslant 1$ for the variables and (ii) an internal loop is inserted within each increment of the transport current, that performs the relaxation of the profiles.

To finish this section, we want to make a further comment about the physical interpretation of the piece-wise approximation. The evolution of the penetration profiles related to the *flux flow* phenomenon takes place connected to the presence of the dissipation term $\Delta t \int_\Omega \mathcal{P}(J)dV$ in the minimisation process. As this term may eventually go to zero, an equilibrium state with circulating critical currents may be reached. Subsequent to the quasi-steady diffusive evolution, the system remains 'static'.

8.6 Finite resistivity: *power-law* approximation

Along the lines of the previous section, here we add a new calculation method that allows to analyse the effect of finite dissipation and the related time relaxation effects. We will consider a model that allows to incorporate the so-called *flux creep* phenomenon [8]. In brief, this consists of the continuous (though slow) jump of flux bundles downhill in the flux gradient, aided by thermal energy. Such a regime may be very relevant in the case of high-T_c superconductors. Thus, the very small values of the coherence length ξ, together with the actual temperatures of interest, promote noticeable thermally induced flux hopping effects. From the point of view of the mathematical model to be used, it should be consistent with the physical condition that, although small, dissipation is always present. Then, by contrast to the above treatment of flux flow effects, the new principle should include a dissipation term that does not fully cancel at any condition. Relaxation will proceed indefinitely though with a smaller rate as times goes on.

Both for its mathematical simplicity and for the possibility of connecting its parameters to the underlying physics, wide consensus exists that this may be nicely described by the so-called *power-law* model introduced by Rhyner [9] that we express by

$$\mathbf{E}(\mathbf{J}) = \rho(J)\mathbf{J}, \quad \rho \equiv \rho_0 \left(\frac{J}{J_c}\right)^{2\nu-2} \tag{8.45}$$

with ρ_0 and ν phenomenological constants whose values are typically adjusted for a given material at a given experimental range. Apparently, the factor ν may be used as a model parameter that allows to describe materials ranging from the ohmic behaviour (for $\nu = 1$ one has $\mathbf{E} = \rho_0\mathbf{J}$) to the critical state regime (for $\nu \gg 1$ one has $E \approx 0$ for $J < J_c$ and $E \to \infty$ for $J > J_c$).

Again, connected to the relation $\mathbf{E} = \nabla_J\mathcal{P}$, one may evaluate the dissipation function to be used in the minimisation algorithms:

$$\mathcal{P} = \frac{\rho_0 J_c^2}{2\nu}\left(\frac{J}{J_c}\right)^{2\nu} \tag{8.46}$$

Let us build up the variational principle for this case and analyse the influence of the power-law exponent on the physical properties deduced. Again, we will concentrate on the transport problem for the slab geometry (as in sections 8.2 and 8.5) so as to allow comparison among the critical state, flux flow and flux creep regimes.

8.6.1 Minimisation

The discretised statement that describes the flux creep regime reads

$$\begin{cases} \texttt{minimise} \left[\dfrac{1}{2}\langle\mathbf{J}|\mathbf{m}|\mathbf{J}\rangle - \langle\mathbf{J}^\vee|\mathbf{m}|\mathbf{J}\rangle + \alpha\langle\mathbf{J}^\nu|\mathbf{J}^\nu\rangle\right] \\[4mm] \texttt{for } \displaystyle\sum_{i=1}^{n} J_i = \dfrac{K}{2J_c d}\, n \end{cases} \tag{8.47}$$

Here we have introduced the constant

$$\alpha = \frac{\rho_0\Delta t}{2\mu_0\nu d^2} \tag{8.48}$$

and, as introduced in chapter 6.1, we use the notation '·' to indicate the element-by-element operation, i.e. $\langle\mathbf{J}^\nu| \equiv (J_1^\nu, J_2^\nu, \ldots, J_n^\nu)$.

8.6.2 Application and analysis

As in the previous section, the application of the minimisation principle with a dissipation term requires a double iteration process. On the one side, the external source contribution enters as a given value of the full transport current (i.e. by updating K in (8.47)). On the other side, having established a value of K one needs to evaluate the relaxation process in which the internal current density profiles evolve in a series of 'substeps' (flow diagram (8.43)). Figure 8.10 displays the reaction of a superconducting sample that obeys the power-law model to the application of a transport current. We show the results for four different values of the exponent ν that characterises the sample. In all cases, we have assumed that a transport current of value $0.25K_c$ is applied, and evaluate the subsequent establishment of the penetration profile. As a general trend, observe that the higher the value of ν, the faster the relaxation process, and the closer that one gets to the ideal critical state behaviour.

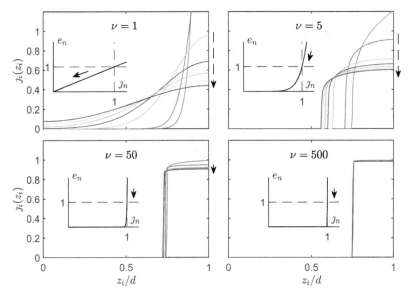

Figure 8.10. Time relaxation of the current density penetration profile in the transport problem for the slab (figure 8.2), according to the power-law model (8.47) for several values of the exponent ν. In all cases, a transport current $K = 0.25K_c$ has been applied. We show the evolution of the subsequent penetration profiles. The insets display the $e(j)$ law that is followed by the outer parts of the slab ($i = n$) 🔗. Source codes available at https://iopscience.iop.org/book/978-0-7503-2711-4.

More in detail:

(i) For the lowest value of ν (ohmic limit $\nu = 1$), as expected, one observes a linear diffusion regime, with $j(z)$ spreading within the sample towards a flat profile, eventually reached in the stationary regime.

(ii) For the highest values of ν (critical state limit), one practically obtains the penetration profile advanced by Bean's model, and the relaxation process takes place in a very small time interval. The current density quickly goes to J_c and the electric field vanishes.

(iii) For intermediate values of ν one observes a relaxation of the current density profile that goes on for current densities below the critical threshold ($J < J_c$). A step-like structure is practically stabilised after some units of time, but for values of $J(z)$ below the model's parameter J_c. Also noticeable is the fact that induced electric fields are predicted at lower values of the current density.

The source codes delivered for the investigation of the power-law model are the following

- `slab_PL_fmincon_transport.m` (main) ; `cost.m`
- `transport.m` (only for OCTAVE users)
- `matrixMXt.m`; `selfXt.m`; `mutualXt.m`
- `relax_PL_profile.m`; `createfigure_PL.m`

Some eventual comments are deserved because, owing to the nature of the problem, the quadratic programming utility employed in the previous sections (8.2 to 8.5) may not be used here. The function to be minimised is no longer quadratic in the variables! Instead, a general purpose minimisation function implemented in MATLAB, i.e. fmincon has been introduced. In OCTAVE the alternative is sqp. In brief, these solvers allow to find the minimum of a wide class of nonlinear multivariable functions with a given set of equality/inequality constraints. The definition of the function to be optimised is done through the accompanying script cost.m Although we prefer to avoid the technical application details (that the reader will find in the form of comments in the attached scripts), we want to mention that the user has the chance to choose between a set of options for the underlying algorithm. Optimally, the selection of such options depends on the actual function to be minimised, that usually is not a trivial task. We have already 'tuned' them up for the power-law dependence in the current example, but it could require further consideration if other situations are to be investigated.

8.7 Review exercises and challenges

8.7.1 Review exercises

Review exercise R8-1

Transform the statement of the transport problem considered in section 8.2 so as to obtain the formulation of the magnetisation problem, i.e. a magnetic field is applied parallel to the x-axis in figure 8.2.

Hint: you will need to reevaluate the mutual induction coefficients so as to adapt to the magnetisation symmetry (shielding condition in figure 8.2), encode a ramp of applied magnetic field and add the corresponding coupling term in the minimisation principle that replaces the transport condition. In order to obtain the elements m_{ij}, start with equation (8.11) and notice that, for this case, summation must be for $j > i$. Then, by squaring the bracket, and grouping terms one gets $m_{ii} = 2(1/4 + i - 1)$, $m_{ij} = 1 + 2\min\{i, j\}$.

Review exercise R8-2

Transform the statement of the problem in section 8.2 into the transport problem for a finite geometry, viz.: superconducting tape of width $2a$ across the x-axis, and negligible thickness d. An imposed current I along y-axis is assumed. This problem has an analytical solution introduced by Brandt and Indenbom [10] in terms of the so-called 'sheet current' K given by

$$K(x) = \begin{cases} \dfrac{2K_c}{\pi}\tan^{-1}\left[\dfrac{a^2 - b^2}{\sqrt{b^2 - x^2}}\right] & |x| \leqslant b \\ K_c & b \leqslant x \leqslant a, \end{cases} \qquad (8.49)$$

with $b = a\sqrt{1 - I^2/(2aK_c)^2}$. Check the numerical solution against this expression. Generalise to the case of ac cycles in the transport current.

Hint: the mutual inductance coefficients for long parallel wires of radius R and centres at the points $(x_i, 0)$ and $(x_j, 0)$ read:

$$M_{ii} = \frac{\mu_0}{8\pi}$$

$$M_{ij} = \frac{\mu_0}{2\pi} \ln \frac{R^2}{(x_i - x_j)^2}$$

It is useful to notice that, because $K(x) = K(-x)$ one may reduce the number of unknowns of the problem by grouping terms in the minimisation principle. Effectively, this means that one may just work in the region $x > 0$ and replace the mutual inductance coefficient: $M(x_i, x_j) \rightarrow M(x_i, x_j) + M(x_i, -x_j)$.

Review exercise R8-3
  Starting from the definition

$$\mathbf{A}(\mathbf{r}) = \frac{\mu_0}{4\pi} \int \frac{\mathbf{J}(\mathbf{r}')}{\|\mathbf{r} - \mathbf{r}'\|} d^3\mathbf{r}' \qquad (8.50)$$

show that the vector potential related function defined in section 8.3 $\psi^{sc} = 2\pi\rho A_{SC}$ may be evaluated in terms of the mutual inductance matrix between the superconducting ith circuits and the 'virtual' circuits at the desired destination points of coordinates (ρ_j, z_j), i.e. $\psi_j^{sc} = \sum_i m_{ij} J_i$, (in matrix form: $|\psi^{sc}\rangle = \mathbf{m}|\mathbf{J}\rangle$).

Hint: discretise the above equation in view of figure 8.4 and apply it to the system formed by the $i = 1, 2, \dots, n$ superconducting circuits and the virtual set of circuits $j = 1, 2, \dots, n'$.

Review exercises R8-4
  Based on section 8.3 evaluate the response of a superconducting long cylinder in the critical state to a uniform magnetic field parallel to its axis. Check the exact formulas in section 9.3.2. Plot (together) the magnetisation hysteresis cycles obtained by both methods.

Hint: according to the results displayed in figure 8.6 one may test some points above and below the aspect ratio $2R/L = 0.33$ so as to establish the validity of the *infinite* cylinder approximation.

Review exercise R8-5

Based on section 8.3 evaluate the response of a superconducting disk in the critical state to a uniform magnetic field. Check the formula by Mikheenko and Kuzovlev [11]

$$
K(\rho) = \begin{cases} \dfrac{2K_c}{\pi} \tan^{-1}\left[\dfrac{(\rho/R)\sqrt{R^2 - a^2}}{\sqrt{a^2 - \rho^2}} \right] & \rho \leqslant a \\ K_c & a \leqslant \rho \leqslant R \end{cases}
\tag{8.51}
$$

Here $a \equiv R/\cosh(2H_0/J_c d)$, H_0 stands for the uniform field applied perpendicular to the disk, d for its (small) thickness, and R its radius. As above K denotes the so-called sheet current.

Hint: use the source codes corresponding to figure 8.5 and modify the script `generate_grid.m` so as to define a single layer of circular circuits with growing radius.

Review exercise R8-6

Use the power-law model for $\nu = 1$ to investigate the magnetic diffusion problem in ohmic material. Compare the time dependent profiles corresponding to the upper/left pane in figure 8.10 to the solution of the diffusion equation for the slab geometry, i.e.

$$
\frac{\partial^2 B_x}{\partial z^2} = \mu_0 \sigma \frac{\partial B_x}{\partial t}
\tag{8.52}
$$

Hint: the above diffusion equation may be solved by means of many available scripts (for example in MATLAB). In particular, as it coincides formally with the heat equation in 1D, one may straightforwardly adapt the available codes to this situation. It will be useful to recall that, corresponding to a given value of transport current per unit length K, the boundary conditions for the magnetic field read $B(z = \pm d) = \pm \mu_0 K/2$.

8.7.2 Challenges

Exercise C8-1

A useful 2D configuration that may be resolved with the tools introduced in this chapter is the so-called 'flux shaking' experiment. This may be realised by applying a uniform transverse field to a long sample, and imposing some variation (left panel) to its components. For instance, according to the geometry in figure 8.1 $\mathbf{H}_{applied} = (H_{x_0}(t), 0, H_{z_0}(t))$. This configuration was considered by Brandt and Mikitik [2]. By means of such model, they could prove that a however weak oscillating *field* applied transverse to the magnetic moment of the sample may lead to the relaxation of this quantity, even though the experimental conditions

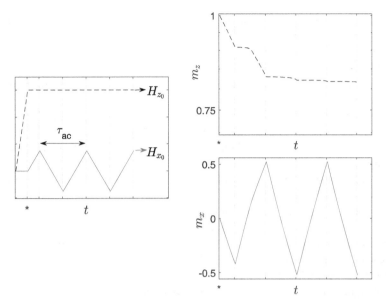

Figure 8.11. Evolution of the magnetic moment components in the proposed *flux shaking experiment*. m_x and m_z are normalised to the value of m_z at the beginning of the ac cycle (marked by '*'). We took a long rod along the y-axis with square cross section.

correspond to the critical state regime. A quasi-1D model was used (long tape), that has been later upgraded to a 2D situation (long rectangular rod) [3]. Here, the reader is invited to simulate such situation by modifying the statement in section 8.3.

For this purpose: (i) define a grid of points (x_i, z_i) that represent the cross section of a long rectangular bar (be mindful about the origin of coordinates), (ii) consider the long-wire approximation for the elementary circuits (as in exercise R8-2) and the corresponding inductance matrix elements, (iii) define the external field process (for instance, ramp up H_z to a definite value, and then make AC cycles with H_x). Write down valid expressions for the associated vector potential, and (iv) evaluate the magnetisation components by applying equation (8.28). In particular, we suggest to check the effect of the AC cycles on $M_z(t)$.

Hint: start by showing that $\mathbf{A}_0(x, t) = \mu_0(0, xH_{z_0} - zH_{x_0}, 0)$ differs by a gradient from $(\mu_0/2)\mathbf{H} \times \mathbf{r}$, and take advantage of this result. It is convenient to impose the condition of vanishing transport current, i.e. $\sum J_i = 0$ explicitly (typical results in figure 8.11).

References

[1] Badía-Majós A and López C 2012 Electromagnetics close beyond the critical state: thermodynamic prospect *Supercond. Sci. Technol.* **25** 104004

[2] Brandt E H and Mikitik G P 2002 Why an ac magnetic field shifts the irreversibility line in type-ii superconductors *Phys. Rev. Lett.* **89** 027002

[3] Badía-Majós A and López C 2007 Critical-state analysis of orthogonal flux interactions in pinned superconductors *Phys. Rev.* B **76** 054504

[4] Peña-Roche J and Badía-Majós A 2016 Modelling toolkit for simulation of maglev devices *Supercond. Sci. Technol.* **30** 014012

[5] Grover F W 1981 *Inductance calculations* (Research Triangle Park, NC: Instrument Society of America)

[6] Brandt E H 1998 Superconductor disks and cylinders in an axial magnetic field. I. flux penetration and magnetization curves *Phys. Rev.* B **58** 6506–22

[7] Sánchez A and Navau C 2001 Magnetic properties of finite superconducting cylinders. I. uniform applied field *Phys. Rev.* B **64** 214506

[8] Tinkham M 1996 Introduction to Superconductivity *International Series in Pure and Applied Physics* (New York: McGraw-Hill)

[9] Rhyner J 1993 Magnetic properties and ac-losses of superconductors with power law current-voltage characteristics *Physica C: Supercond. Appl.* **212** 292–300

[10] Brandt E H and Indenbom M 1993 Type-II superconductor strip with current in a perpendicular magnetic field *Phys. Rev.* B **48** 12893–906

[11] Mikheenko P N and Kuzovlev Y E 1993 Inductance measurements of HTSC films with high critical currents *Physica C: Supercond. Appl.* **204** 229–36

Macroscopic Superconducting Phenomena
An interactive guide
Antonio Badía-Majós

Chapter 9

Shape effects: demagnetising fields

In the research of the magnetic properties of superconductors one frequently faces the question of shape effects. This topic is certainly very well known in the field of magnetism, and was already present in the classical treatise by J C Maxwell. In brief, along the course of analysing the response of a given sample to some excitation (applied magnetic field) one obtains a 'convolution' of (i) the intrinsic material's properties, (ii) the actual excitation, and noticeably (iii) the sample's geometry. It is only for extreme situations (long cylinders under parallel magnetic field, as in figure 9.1) that one may include the geometrical aspects with ease. From the experimental point of view, this is a very relevant question because measurements on a sample of arbitrary shape may be useless if the correct 'deconvolution' is not performed. Thus, as in many instances one may not shape the sample as a long cylinder, some tool for analysing real measurements is needed. In this chapter we offer some resources dedicated to the study of geometries that although simple may help in many real life applications. Especially, we concentrate on the (perfect) Meissner state and critical state properties of cylinders and ellipsoids of revolution under uniform magnetic fields along the axis. Other geometries such as those of rectangular cross sections are sometimes approximated by inscribing in the ellipsoidal case, or may be studied by specialised methods as those presented in chapter 10 for thin samples in perpendicular magnetic field.

9.1 Statement of the problem

The topic of shape effects in superconductors may be introduced with the help of figure 9.1. Let us assume that a superconducting cylinder of arbitrary cross section is under the effect of a uniform magnetic field $\mu_0 \mathbf{H}_a$ applied along the axis (say z-axis). In the case of low fields ($H_a < H_{c1}$ for type-II materials) and assuming that the penetration depth λ_L is much smaller than the typical sample's dimensions, one may describe the response of the sample in terms of a surface current distribution $\mathbf{K}$ flowing along the lateral wall, that produces a magnetic field $-\mu_0 \mathbf{H}_a$ within the

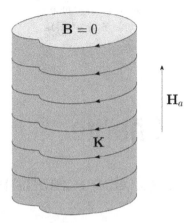

Figure 9.1. Sketch of the expulsion of magnetic field (Meissner state) in a long cylinder with arbitrary cross section.

superconductor. If the cylinder is very long, its response will be well described by the 'infinite-cylinder' approximation[1]. This means that **K** does not change along the axis and one may use the infinite solenoid model. The application of Ampère's law straightforwardly leads to obtaining the magnetic field created by the superconductor in terms of **K**. In fact $\mathbf{B}_{sc} = \mu_0 K \hat{k}$ and then[2]

$$\mathbf{B}_{full} = \mu_0 H_a \hat{k} + \mu_0 K \hat{k} = 0 \implies K = -H_a \tag{9.1}$$

On the other hand, the z-component (and unique) of the magnetic moment per unit volume of such cylinder would be given by

$$M_v = \frac{K \cdot L \cdot A}{A \cdot L} = K = -H_a \tag{9.2}$$

with A, L being the cross-sectional area and length of the cylinder, respectively.

Thus, the 'ideal' linear magnetic susceptibility of the sample would be

$$\chi_{ideal} = \frac{M_v}{H_a} = -1 \tag{9.3}$$

Let us now imagine that the cylinder is not so long, and try to foretell what will occur. If one exploits further the analogy to the finite solenoid [1], the configuration with $K = -H_a$ would not cancel the full magnetic field but approximately in a small region around the centre[3]. Thus, more current is needed close to the edges, and if a solution exists that gives $B = 0$ all within the cylinder, this leads to values of χ smaller

---

[1] Further in this chapter we provide the tools for validating such approximation in real cases.
[2] In this equation **K** is oriented according to $\hat{\imath} \times \hat{\jmath}$.
[3] We recall that such field reaches (at most) a maximum value $\mu_0 K$ at the centre and decays toward the edges of the solenoid as may be deduced from the familiar expression $\mu_0 K (\cos \theta_2 - \cos \theta_1)/2$.

than -1 (M_v more negative)[4]. In fact, for the case of ellipsoidal samples, one may show that the magnetic susceptibility in the Meissner state may be expressed [2]

$$\chi = \frac{-1}{1 - N} \qquad (9.4)$$

with N, the so-called *demagnetising factor*, being a geometry dependent parameter that is tabulated in many sources. For instance, the reader is addressed to the classical papers by Stoner [3] and Osborn [4]. The basic idea behind such concept comes from the theory of magnetism. When a paramagnetic ellipsoid is subjected to a uniform external field $\mathbf{H}_a$, the above mentioned edge effect produces a so-called *demagnetising field* $\mathbf{H}_d$ within the sample, given by $\mathbf{H}_d = -N\mathbf{M}_V$. In general, N is a tensor, but keeping within the situation of applying the field along the ellipsoid z-axis it is a scalar and one may straightforwardly obtain the related magnetic susceptibility

$$M_v = \chi_{\text{ideal}}(H_a + H_d) = \chi_{\text{ideal}}(H_a - N\ M_v) \qquad (9.5)$$

Then, adopting the 'experimental' point of view that relates M_v to the applied magnetic field $M_v \equiv \chi H_a$ one may derive

$$\chi = \frac{\chi_{\text{ideal}}}{1 + N\chi_{\text{ideal}}} \qquad (9.6)$$

from which one obtains equation (9.4) for the perfectly diamagnetic case.

In the theory of magnetism the concept of demagnetising fields has been very useful, but strictly applies to the case of ellipsoidal bodies. For non-ellipsoidal samples, neither $\mathbf{H}_d$ nor $\mathbf{M}$ are uniform within the sample, and different approaches are used to deal with the related effects. Currently, most of them rely on the numerical solution of the full problem, which consist of directly calculating the sample's magnetic moment through the constitutive equations under specific boundary conditions for the geometry of interest. This will be the direction of this chapter for the case of superconductors. Some peculiarities deserve to be mentioned in advance.

- For the ideal Meissner state, the induced supercurrents produce a uniform field within the sample, that exactly cancels the applied field $\mathbf{H}_a$ regardless of the shape and the direction of $\mathbf{H}_a$.
- For the critical state, the induced supercurrents do not produce a uniform field within the sample under any condition

These properties, as well as their related consequences, will be studied with detail in the forthcoming sections.

---

[4] The exact values are analysed in the forthcoming section 9.2.

9.2 The Meissner state in finite samples: ellipsoids and cylinders

The Meissner state of revolution ellipsoids and circular cylinders under uniform axial field may be studied by recalling the minimisation principle stated in equation (2.9) combined with the finite element technique introduced in chapter 8. To be specific, this may be implemented just by simplifying equation (8.26): (i) deleting the terms that couple to the previous time layer and (ii) relaxing the constraint for the current density and allowing this variable to take whatever values necessary to minimise the energy. Alternatively, a rather more efficient method is suggested. To start with, one may restrict the minimisation process to a surface layer and avoid dealing with the interior points. Also, noticing that the minimisation of a non-constrained quadratic function admits an explicit solution, one may just apply a matrix multiplication so as to obtain the solution. Thus, in matrix form, and using the notation $|\mathbf{K}\rangle$ for the finite element vector that stores the Meissner currents at the points of the surface, one gets the principle

$$\texttt{minimise} \quad \mathbf{U} = \frac{1}{2}\langle\mathbf{K}|\mathbf{m}|\mathbf{K}\rangle + \langle\psi_{\text{Applied}}|\mathbf{K}\rangle \tag{9.7}$$

whose solution in matrix form is just[5]

$$|\mathbf{K}\rangle = -\mathbf{m}^{-1}|\psi_{\text{Applied}}\rangle \tag{9.8}$$

Figures 9.2 and 9.3 have been obtained by applying the above equation, which is efficiently dealt with by using MATLAB scripts. As said before, we consider the

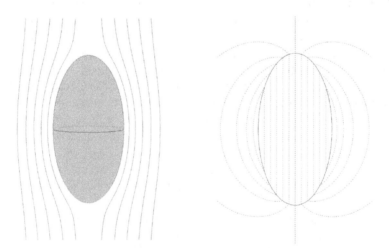

Figure 9.2. Expulsion of magnetic field (Meissner state) for an ellipsoid of revolution (aspect ratio $\zeta = 2$). To the right we show the field lines corresponding to the induced supercurrents. 🔗. Source codes available at https://iopscience.iop.org/book/978-0-7503-2711-4.

---

[5] Recall that $\mathbf{m}$ is the mutual inductance matrix for the superconducting circuits (circular loops in this case) and ψ_{Applied} relates to the vector potential for the applied magnetic field.

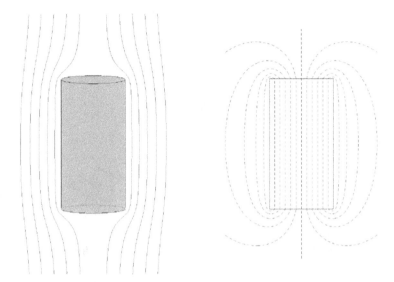

Figure 9.3. Expulsion of magnetic field (Meissner state) in a superconducting cylinder. To the right we show the field lines corresponding to the induced supercurrents. 🔗. Source codes available at https://iopscience.iop. org/book/978-0-7503-2711-4.

ellipsoidal geometry for the particular case of revolution symmetry. Thus, using a, b, c for the semi-axes along $\hat{\imath}, \hat{\jmath}, \hat{k}$, we have $a = b$ and define the aspect ratio $\zeta \equiv c/a$. Under such conditions and for $\mathbf{H}_a = H_a \hat{k}$, supercurrents flow as concentrical circular loops and the matrix elements $\mathbf{m}_{ij}$ are evaluated through the expressions (8.20) and (8.21). One may notice that, as expected, the Meissner state leads to flux expulsion within the sample. This condition is reached because supercurrents adopt such a profile that they create an opposing magnetic field that exactly cancels the applied excitation. This is emphasised to the right part of the figures, which shows the uniform field created by the superconductor within the sample. We stress that this property is valid for any geometry and it is allowed by the absence of restrictions for the circulating supercurrents.

To end this section, we include a plot of the sample's magnetisation (full magnetic moment $\boldsymbol{\mu}$ as defined in equation (8.28) divided by the volume) obtained for ellipsoids of various aspect ratios. As expected, all the curves display a linear dependence on the magnetic field[6], with increasing slope as ζ diminishes. In fact, one may easily check that the susceptibility obtained by dividing over volume and applied magnetic field perfectly coincides with the value calculated via equation (9.4) with the demagnetising factor N taken from the reference tables. A very similar behaviour may be observed for circular cylinders, as the reader may check in exercise R9-2. However, we want to stress that for such case, the concept of demagnetising fields and factors is more complex. To our knowledge, the best

---

[6] Both the magnetic moment of a certain current distribution and the created magnetic field are multiplied by the same scale factor when the full current distribution is multiplied by some value.

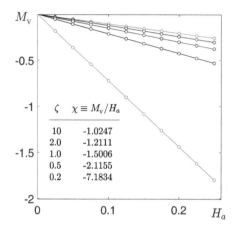

Figure 9.4. Volume magnetisation for ellipsoidal samples of different aspect ratios and related magnetic susceptibility obtained as the slope of the lines. 🔗. Source codes available at https://iopscience.iop.org/book/978-0-7503-2711-4.

approximation to the real facts through an expression like equation (9.4) is achieved through the so-called *magnetometric demagnetising factors* calculated by Chen, Brug and Goldfarb [5]. Figures 9.2, 9.3, and 9.4 have been compiled with the following scripts:

- `ellip_meiss.m` (main for figure 9.2)
- `cyl_meiss.m` (main for figure 9.3)
- `generate_grid_surf_ellip.m` (for figure 9.2)
- `generate_grid_surf.m` (for figure 9.3)
- `matrixM.m`; `self.m`; `mutual.m`
- `psi_pot_unif.m`
- `psi_pot_SC.m`
- `plot_meiss_ellips.m` (for figure 9.2)
- `plot_meiss_cyl.m` (for figure 9.3)
- `z_ellipsoid.m` (for figure 9.2)
- `e_cylinder.m` (for figure 9.3)
- `magnetisation_field.m` (for figure 9.4)

9.3 The critical state in finite samples

Let us now analyse the influence of shape effects in the magnetic response of samples in the critical state. As in chapter 8, it is meant that one works with hard superconductors in the range of fields well above H_{c1} and well below H_{c2}. We will also concentrate on the response of revolution ellipsoids and cylinders for uniform fields applied along the axis, i.e. $\mathbf{H}_a = H_a \hat{k}$.

9.3.1 Bean's model in ellipsoids and cylinders

We will use the tools introduced in section 8.3, i.e. the minimisation procedure given by equation (8.26) that allows to obtain the supercurrent distribution $|\mathbf{J}\rangle$ in the finite element approximation shown in figure 8.4. No additional comment is necessary for the case of cylinders with different aspect ratios ζ, but an important remark must be made for the case of ellipsoids. As shown in figure 9.5, owing to the curvature of the boundary, one must design a dedicated strategy in the discretised statement. Ellipsoidal areas may not be exactly covered by a square mesh but only approximated to a given degree of precision by using an increasingly refined mesh. This may be a big handicap for extreme values of ζ. A solution to this problem has been implemented in the provided scripts and is sketched in the figure. Thus, taking advantage of MATLAB's built-in functions `polyshape()`, `intersect()` and `area()`, one may proceed as follows: (i) start with a square mesh of points P_i inscribed in the elliptical contour, (ii) calculate the intersection areas A_i of the elementary cells around the points of the mesh and the ellipse, and (iii) incorporate the distribution of areas to the calculation of the current along each individual circuit: $I_i = j_i A_i$.[7]

The resulting magnetisation profiles for ellipsoids with aspect ratios in the range $0.1 < \zeta < 10$ are shown in figure 9.6. Recall the units that have been used in this plot. The applied magnetic field is scaled by *Bean's penetration field* $H_p = J_c a$ and magnetisation by the saturation value for the ellipsoid $M^* \equiv 3\pi J_c a/32$ (see exercise R9-6). The effect of geometry is visible: as ζ diminishes (the flatter the sample), the hysteresis loop gets more rectangular. To the right of the plot, we show the structure of the magnetic field created by the superconductor for the so-called partial penetration regime. We notice the existence of a uniform internal field for a region within the ellipsoid, the flux free core Ω_0, that coincides with the non-penetrated part ($j_i = 0$, $\forall P_i \in \Omega_0$).

Figure 9.7 shows the magnetisation loops obtained for cylindrical samples with aspect ratios in the range $0.1 < \zeta < 10$. The general features already described for ellipsoids are valid. However, a number of quantitative details related to the actual geometry are to be mentioned. In this case, one has $H_p = J_c R$ and the saturation of the magnetisation takes place for the value $M_p = J_c R/3$, with R being the radius of the cylinder, as mentioned in exercise R9-5. On the other hand, one may notice that, although similar, the shape of the flux free core is different and also appreciate subtle differences in the magnetisation for the partial penetration regime. For both cases, we notice that, by contrast to the case of the Meissner state, the existence of a region

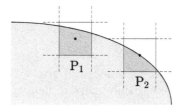

Figure 9.5. Detail of the finite element approximation for the elementary current loops (cross section) in the ellipsoidal geometry.

---

[7] Notice that, for the case of cylinders, the values A_i are just a constant and, thus a common factor in summations. This was used in chapter 8, with the notation $A_i \equiv \delta^2$.

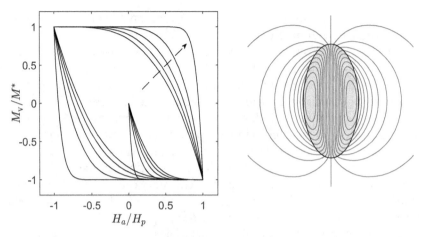

Figure 9.6. Magnetisation loops corresponding to the critical state for ellipsoidal samples with aspect ratios $\zeta = 10, 2, 1, 0.5, 0.1$ (as indicated by the arrow). To the right, we show the magnetic field created by the superconductor in the partially penetrated regime for the case $\zeta = 2$, $H_a = 0.5H_p$. $\mathscr{O}$. Source codes available at https://iopscience.iop.org/book/978-0-7503-2711-4.

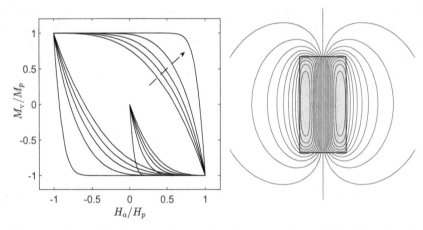

Figure 9.7. Same as figure 9.6 but for cylindrical samples. $\mathscr{O}$. Source codes available at https://iopscience.iop.org/book/978-0-7503-2711-4.

with uniform internal magnetic field does not give way to a linear magnetisation. The reason is that the superconducting current density profile (shaded area in the plots) changes along the process.

Figures 9.6 and 9.7 have been compiled with the following scripts

- `ellip_cs_qp.m` (main for figure 9.6)
- `cyl_cs_qp.m` (main for figure 9.7)
- `generate_grid_ellip_sym_areas.m` (for figure 9.6)

- `generate_grid.m` (for figure 9.7)
- `matrixM.m`; `self.m`; `mutual.m` (`matrixM_sym.m` for figure 9.6)
- `psi_pot_unif.m`
- `psi_pot_ellips_SC_areas.m` (for figure 9.6)
- `psi_pot_cylinder_SC.m` (for figure 9.7)
- `plot_cs_ellip.m` (for figure 9.6)
- `plot_cs_cylinder.m` (for figure 9.7)
- `magnetisation_loop_ellip.m` (for figure 9.6)
- `magnetisation_loop_cylinder.m` (for figure 9.7).

A technical detail to be mentioned here is that, taking advantage of the problem's symmetry, one may pose the finite element formulation in terms of a reduced number of variables. Just recall that, as one could expect, in the configurations studied above the solution is symmetric respect to the horizontal half-plane. Then, one can in fact work with half the number of 'circuits' in the problem. This property has been used in the ellipsoidal geometry, so as to allow a higher resolution without increasing the numeric load (implemented in `generate_grid_ellip_sym_ar-eas.m` and `matrixM_sym.m`). Further details about the method may be found in chapters 10 and 14 (sections 10.1.3 and 14.2.2) in which reductions to one fourth of the original size are applied.

9.3.2 Analytical approximations for extreme geometries

In many experimental studies it is argued that, owing to the either big or small value of the sample's aspect ratio, one may safely apply analytical approximations in order to characterise the response to a given external excitation. In this section, we provide some tools for evaluating the validity of such hypothesis by checking against numerical calculations. As shown in figure 9.8, we consider the limiting cases $\zeta \gg 1$ and $\zeta \ll 1$ for cylindrical samples. Let us analyse them separately.

Long sample limit
According to the left part of the plot one can ascertain that Bean's model for ideally infinite cylinders may be safely applied for $\zeta = 20$.[8] Here, we plot the numeric solution for the actual cylinder and compare to the analytical expressions[9] (as adapted from [6]):

$$
\frac{M_v}{M_p} = \begin{cases} -3\dfrac{H_a}{H_p} + 3\left(\dfrac{H_a}{H_p}\right)^2 - \left(\dfrac{H_a}{H_p}\right)^3; & H_a \leqslant H_p \\ -1; & H_a > H_p \end{cases} \qquad (9.9)
$$

---

[8] The reader is invited to use the provided software to check how this works for other values of ζ.
[9] Equations (9.9) and (9.10) are the cylindrical symmetry counterparts of (3.20) and (3.21) given in chapter 3 for the slab approximation.

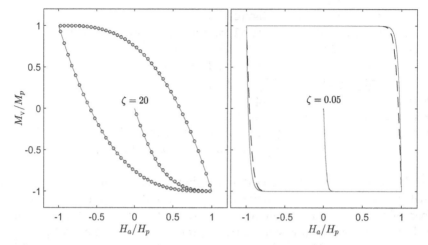

Figure 9.8. Magnetisation loops corresponding to the critical state for cylindrical samples with 'extreme' aspect ratios: $\zeta = 20$ to the left (numeric: symbols) and $\zeta = 0.05$ to the right (numeric: dashed line). For comparison, the continuous lines display the analytical limits (Bean's model, left and Mikheenko's model, right). 🔗. Source codes available at https://iopscience.iop.org/book/978-0-7503-2711-4.

for the initial branch of the cycle, and

$$
\frac{M_{\mathrm{v}}}{M_p} = -\frac{H_a}{H_p} + \frac{H_a H_{\mathrm{M}}}{H_p^2} \pm \frac{1}{2}\left[\left(\frac{H_a}{H_p}\right)^2 - \left(\frac{H_{\mathrm{M}}}{H_p}\right)^2\right]
$$
$$
\pm \frac{1}{4}\left[\left(\frac{H_{\mathrm{M}}}{H_p}\right)^3 + \frac{H_a H_{\mathrm{M}}^2}{H_p^3} - \frac{H_{\mathrm{M}} H_a^2}{H_p^3} + \left(\frac{H_a}{H_p}\right)^3\right]
\tag{9.10}
$$

for the upper/lower courses.

In this case, we have used the relation $H_{\mathrm{M}} = H_p$ for the maximum applied field. H_p and M_p correspond to the quantities defined in the previous section.

Flat sample limit
According to the right panel of figure 9.8, Mikheenko's and Kuzovlev's model [7] for thin discs gives a very reasonable approximation for $\zeta = 0.05$. The numerical results are compared to the magnetisation obtained by integrating the expression for the surface supercurrent on the disk[10], that is the basic quantity of the model. Thus, on the one side, one has

---

[10] Recall the definition of the surface currents for a thin sample $K(r) = \int J(r, z)dz$.

$$K_\varphi(r) = \begin{cases} -\dfrac{2K_c}{\pi} \tan^{-1}\left[\dfrac{(r/R)\sqrt{R^2 - a^2}}{\sqrt{a^2 - r^2}}\right]; & r < a \\ -K_c & ; & a \leqslant r \leqslant R \end{cases} \qquad (9.11)$$

for the initial magnetisation of the disk. Here we have used the definitions

$$K_c \equiv J_c d$$
$$a \equiv R/\cosh(2H_a/K_c)$$

with R, d being its radius and thickness, respectively.

On the other side, for the upper/lower courses of the loop we use

$$K_\varphi(r) = \begin{cases} -\dfrac{2K_c}{\pi} \tan^{-1}\left[\dfrac{(r/R)\sqrt{R^2 - a_M^2}}{\sqrt{a_M^2 - r^2}}\right]; & r < a_M \\ -K_c & ; & a_M \leqslant r \leqslant b \\ K_c & ; & b \leqslant r \leqslant R \end{cases} \qquad (9.12)$$

with

$$a_M \equiv R/\cosh(2H_M/K_c)$$
$$b \equiv \cosh[(H_a + H_M)/K_c]/\cosh(2H_M/K_c)$$

Integration of $K(r)$ gives M_v in the upper course: $H_a(\searrow)$: $H_M \to -H_M$, and the returning branch is obtained by symmetry

$$H_a(\nearrow) = -H_a(\searrow); \qquad M_V(\nearrow) = -M_V(\searrow) \qquad (9.13)$$

Equations (9.11) and (9.12) have been very useful in the investigation of superconducting thin films. In order to gain further insight in the physical problem and practice with the numerical methods the reader is invited to work out exercise C9-1. There, we propose to compare the exact critical state solution that depends on the details across the sample's thickness with Mikheenko's proposal that averages across such dimension. A side benefit of this exercise is that one may use it as a link for going into more complex problems in thin samples as those treated in chapter 10.

Figure 9.8 has been created with the help of the scripts for figure 9.7 and the additional scripts that evaluate the expressions in equations (9.11) to (9.13).

- `bean_model.m`
- `mikheenko_model.m`
- `figure_9_8.m`

9.3.3 The internal magnetic field: $J_c(B)$ versus $J_c(\mu_0 H_a)$

As emphasised by Sánchez and Navau [8], a very relevant consequence of the influence of shape effects in the experimental studies of high-T_c superconductors concerns the determination of the sample's critical current density J_c. In brief, as we explained in sections 3.3 and 4.2, this key parameter is frequently determined from

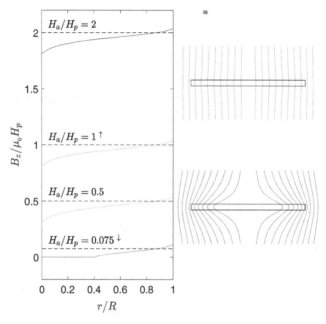

Figure 9.9. Internal magnetic field profiles B_z in the midplane of a thin superconducting cylinder ($\zeta = 0.05$) for different values of the increasing applied field H_a (dashed lines) along the z-axis. To the right, we plot the full magnetic field lines for two values of the applied field, as marked. 🔗. Source codes available at https://iopscience.iop.org/book/978-0-7503-2711-4.

magnetisation measurements, more specifically from the width of the magnetisation loops $\Delta M_v(H_a)$. The problem arises because, indeed, J_c depends on the magnetic field within the sample (actually both on its modulus and orientation). As said in chapter 3, the ansatz of Bean's model (J_c = constant) is a simplifying approximation, whose validity has to be assessed in practice. In [8] it was shown that, for a given material, the best approximation to the actual $J_c(B)$ may be obtained from magnetisation measurements in flat samples with the magnetic field applied along the shortest dimension. The authors give quantitative criteria for the consistency of results for different assumptions on the underlying $J_c[B(\mathbf{r})]$ dependence. Here, we want to recall that, as explained in section 2.6, within our macroscopic theory, $\mathbf{B(r)}$ is the full magnetic field within the sample for a resolution that smoothes out the fluctuations due to individual vortices. It is contributed by the macroscopic supercurrents, as well as by the external sources, i.e. $\mathbf{B(r)} = \mathbf{B}_{sc}(\mathbf{r}) + \mu_0 \mathbf{H}_a$, and only $\mathbf{H}_a$ is at hand in the experiment[11]. Here, we will give a heuristic argument to justify the result in [8]. Figure 9.9 shows the internal magnetic field profiles for a thin superconductor, assuming a field independent critical current density. It is apparent that the difference between the applied magnetic field and the internal field becomes a small factor as the typical value $\mu_0 H_p$ is reached. In other words, although always

---

[11] Though strictly unnecessary when the equilibrium magnetisation effects may be neglected, some authors introduce the definition of the magnetic field intensities $\mathbf{H} \equiv \mathbf{B}/\mu_0$ and $\mathbf{H}_{sc} \equiv \mathbf{B}_{sc}/\mu_0$.

present, the field B_{sc} becomes negligible as the sample goes thinner. Then one may safely use

$$\Delta M(B) \approx \Delta M(\mu_0 H_a) \qquad (9.14)$$

Figure 9.9 was obtained with the help of the software for figure 9.7 and the additional script figure_9_9.m Further practice with the inversion problem related to specific $J_c(B)$ dependencies is suggested in exercise C9-2.

9.4 Review exercises and challenges

9.4.1 Review exercises

Review exercise R9-1

Reproduce figures 9.2 and 9.3 for the case of a finite London's penetration depth λ_L. In order to find noticeable effects λ_L must be comparable to the sample's typical dimensions.

Hint: modify the related software, according to the application of equation (2.9).

Review exercise R9-2

Reproduce figure 9.4 with related table for the case of finite cylinders and check the results against the *magnetometric* demagnetising factors in [5].

Hint: straightforwardly use the cylindrical geometry instead of ellipsoidal in the related scripts.

Review exercise R9-3

Obtain the exact expression for the full penetration field of a cylindrical super-conductor in the critical state if the applied field is uniform and along the axis of the cylinder:

$$H_\zeta = H_p \, \zeta \, \sinh^{-1}\left(\frac{1}{\zeta}\right) \qquad (9.15)$$

with $\zeta \equiv L/2R$ being the cylinder's aspect ratio and $H_p \equiv J_c R$ the approximation for large values of ζ. Assume that $J_c = $ constant.

Hint: consider a cylinder fully penetrated by circular current loops with a density J_c and evaluate the field created by such currents at the centre. It may be useful to treat the cylinder as a superposition of hollow cylinders with uniform surface current densities.

Review exercise R9-4

Obtain the exact expression for the full penetration field of a spherical superconductor in the critical state if the applied field is uniform:

$$H^* = \frac{\pi}{4} J_c R \qquad (9.16)$$

with R being the radius of the sphere. Assume that J_c = constant.

Hint: it may be useful to describe the fully penetrated sphere as the superposition of spherical shells with surface currents of uniform distribution.

Review exercise R9-5

Show that the saturation magnetisation (magnetic moment per unit volume) of a cylindrical superconductor in the critical state under parallel field is given by

$$M_p = \frac{J_c R}{3} \qquad (9.17)$$

i.e. depends on radius but not on length. Assume that J_c = constant.

Hint: recall that the superconductor may be described by a collection of circular loops of current with constant density J_c, and integrate over the cylinder to obtain the magnetic moment.

Review exercise R9-6

Show that the saturation magnetisation (magnetic moment per unit volume) of an ellipsoid of revolution in the critical state under applied field along its axis of revolution is given by[12]

$$M^* = \frac{3\pi}{32} J_c\, a \qquad (9.18)$$

with a being the semiaxis of revolution. It is independent of the other semiaxis (say c) Assume that J_c = constant.

Hint: see the exercise above.

---

[12] A number of useful expressions and related discussion concerning the penetration of magnetic flux in ellipsoidal samples in the critical state may be found in the work by Navarro and Campbell [9].

9.4.2 Challenges

Exercise C9-1

Consider the thin superconducting cylinder studied in section 9.3.2 ($\zeta = 0.05$) and:
- (a) For an applied field of value $H_a = 0.075H_p$ make a plot of the surface current density as a function of r by using equation (9.11)
- (b) In such conditions, and using the numerical method for the critical state in cylinders, calculate the volume current density profile and make a plot in the fashion of figure 9.7
- (c) From the above calculated current density distribution make a plot of the magnetic field lines (in the fashion of figure 9.9)
- (d) By comparing (a), (b) and (c), make a physical interpretation of Mikheenko's model and related quantities, i.e. $B_z(r)$, $K(r)$, a.

Hint: see figure 9.10.

Exercise C9-2

Reproduce the results in [8] for the inversion of $J_c(B)$ from experimental magnetisation loops $\Delta M(\mu_0 H_a)$, i.e. reproduce figure 4 in that paper.

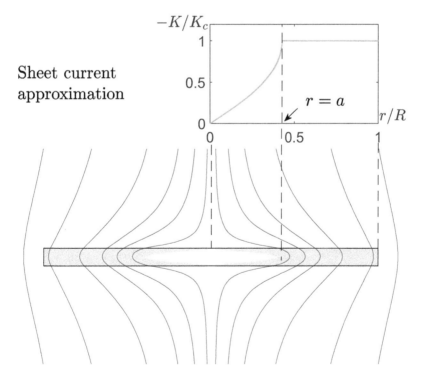

Figure 9.10. Comparison of Mikheenko's formula for the sheet current in thin sample's (equation (9.11)) and the exact 3D penetration profile of the current density in a thin cylinder ($\zeta = 0.05$).

Hint: consider the scripts related to figure 9.7 and replace the inequality condition $J \leqslant J_c$ by $J \leqslant J_c(b)$ in `cyl_cs_qp.m` with B calculated from the previous step. Use small intervals ΔB. According to figure 9.9, B may be approximated by $\mu_0 H_0$ in the range of interest (fully penetrated state).

References

[1] Jackson J D 1975 *Classical Electrodynamics* 2nd edn (New York: Wiley)
[2] Cape J A and Zimmerman J M 1967 Magnetization of ellipsoidal superconductors *Phys. Rev.* **153** 416–21
[3] Stoner E C 1945 The demagnetizing factor for ellipsoids *Philos. Mag.* **36** 803–21
[4] Osborn J A 1945 Demagnetizing factors of the general ellipsoid *Phys. Rev.* **67** 351–7
[5] Chen D X, Brug J A and Goldfarb R B 1991 Demagnetizing factors for cylinders *IEEE Trans. Magn.* **27** 3601–19
[6] Bean C P 1964 Magnetization of high-field superconductors *Rev. Mod. Phys.* **36** 31–9
[7] Mikheenko P N and Kuzovlev Y E 1993 Inductance measurements of htsc films with high critical currents *Physica C: Supercond. Appl.* **204** 229–36
[8] Sánchez A and Navau C 2001 Critical-current density from magnetization loops of finite high-tc superconductors *Supercond. Sci. Technol.* **14** 444–7
[9] Navarro R and Campbell L J 1991 Magnetic-flux profiles of high-T_c superconducting granules: Three-dimensional critical-state-model approximation *Phys. Rev.* B **44** 10146–57

IOP Publishing

Macroscopic Superconducting Phenomena
An interactive guide
Antonio Badía-Majós

Chapter 10

Thin superconductors: the stream function method

Not by chance, in a good number of fundamental studies as well as of applications, superconducting samples are in the form of thin films. Micro- and nanofabrication techniques allow to produce extremely high-quality tailored samples with fine control of their microstructure. Additionally, the possibility of reducing the thickness even to a few unit cells of the material allows to observe unsuspected new phenomena. In this chapter, we will concentrate on some tools for understanding the response of thin samples in the basic configuration of figure 10.1. In brief, we deal with thin platelets, i.e. $a \gg d$ that may be treated as quasi-2D in the macroscopic scale. In particular, this involves the characteristic length scale of the applied magnetic field, and means that variations across the thickness of the sample may be neglected. Nevertheless, concerning the superconducting material, we will still assume that d is not small as compared to the characteristic lengths, mainly concerning ξ so as to avoid fundamental modifications of the underlying physics.

The theory presented below applies, for instance, to HTS (high-temperature superconducting) films with submicron thickness ($d \lesssim 1\ \mu m$) and typical dimensions in the mm range, i.e. $a \lesssim 5$ mm. In this case, and for applied magnetic fields in the range of 1 Oe ($\mu_0 H_a = 10^{-4}$ T), one may assume that the critical state approximation revised in chapter 9 may be used. In fact, as recalled by Mikheenko and Kuzovlev [1], an eventual consideration should be done in order to justify the theory. As in thin samples, the fields decay with Pearl's [2] effective length $\lambda_{\text{eff}} = \lambda_{\text{L}}^2/d$, the formation of the critical state (intervortex distances are negligible) requires the supplementary condition[1]: $a \gg \lambda_{\text{L}}^2/d$.

The basic framework to solve critical state problems in the thin sample limit will be the variational statement exploited in chapters 8 and 9. In fact, one could straightforwardly use the method introduced in section 8.3 to investigate the thin

---

[1] The set of conditions mentioned above is typically met in magnetisation experiments with HTS thin films [1].

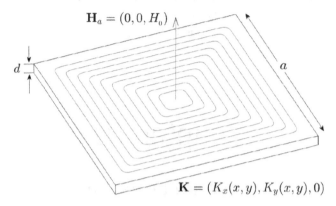

$$\mathbf{H}_a = (0, 0, H_0)$$

$$d$$

$$a$$

$$\mathbf{K} = (K_x(x, y), K_y(x, y), 0)$$

Figure 10.1. Superconducting square platelet. Shown are the (2D) current density streamlines induced by applying a field perpendicular to the platelet. Ideally $a \gg d$.

disk limit as a particular case of the superconducting cylinder under uniform magnetic field. However, there is a fundamental 'simplification' of the problem that was implicitly applied in such case. By symmetry, one could assume that the superconducting currents follow circular loops. In practice, under such condition, the variational statement in equation (8.26) can be solved because the mutual inductance matrix coupling the loops is given by (8.20). But, what if one wants to solve the critical state problem in a square platelet like that in figure 10.1? Certainly, the current loops exist, as do the M_{ij} elements, but they are unknown. One must definitely search for some new method. A widely used strategy is described in this chapter and relies on the so-called *stream function formalism*.

The essentials of this theory may be synthesised as follows.

- From the physical side, the condition that the current density must flow in the form of current loops may be formulated by a 'potential-like' theory. One may describe the 2D current density $\mathbf{K}$ in terms of the gradient of a scalar function $\sigma(x, y)$, the *stream function*.
- By using finite differences, one may obtain a variational principle fully equivalent to (8.26) in terms of the unknown $\sigma(x, y)$ and a geometrical matrix that generalises $\mathbf{m}$. Having solved for σ one may straightforwardly obtain $\mathbf{K}$, and thus the actual circuits of current are ultimately determined.
- From the technical side, in order to obtain a reasonable output of the numerical procedure, one must use a big number of unknowns, i.e. σ must be evaluated in a big number of points (x_i, y_i). In order to minorate the slowdown or hindrance of the numerical procedure, and based on symmetry arguments, we will introduce a technique for reducing of the problem's size without loss of physical information.

10.1 Statement of the problem

As indicated in figure 10.1, our purpose is to find the distribution of current density that is induced in a superconducting platelet lying on the XY plane, when an external magnetic field is applied. Only for illustration, the figure shows the case of a uniform field along the Z-axis of a square platelet but, as will be seen below, the formulation may be applied in rather general conditions as concerns the applied field and shape of the platelet.

Along the lines of chapters 8 and 9, here we develop a numerical method based on the variational statement of the problem. The general ideas for using variational formulations in thin film magnetisation problems were established by L Prigozhin and may be found in [3]. On the other hand, for the readers' sake we mention that analytical approximations providing solutions for the critical state in planar samples are available in the literature. Some relevant contributions may be found in [1, 4, 5], and a thorough review of the state of the art in [6]. In general, such methods rely on the solution of some integral equation through more or less sophisticated methods.

10.1.1 The stream function formalism

Inherited from classical fluid dynamics, the stream function method applies to such problems in which some physical quantity flows in a closed region in the absence of sources or sinks. In our case, such a quantity is the electrical current density $\mathbf{J}$. In fact, owing to the planar geometry, we use the *sheet current* $\mathbf{K}(x,y)$, i.e.

$$\mathbf{K}(x,y) = \int_{-d/2}^{d/2} \mathbf{J}(x,y,z)dz \qquad (10.1)$$

Here d is the thickness of our platelet.

Recall that the condition of charge conservation in the quasistatic regime (neither accumulation, input or output of net charge occur) reads $\nabla \cdot \mathbf{K} = 0$. In practice, this may be included in the theory in different ways. For instance, being focused on numerical methods, we mention that computational electromagnetism is equipped with sophisticated techniques (vector finite elements) that allow to handle such problems even in 3D situations. The interested reader may find an excellent introduction to such methods in the book by A Bossavit [7]. Here, being focused on the straightforward solution of 2D problems we present a method based on the discretisation of a *scalar* function directly related to $\mathbf{K}$.

To start with, notice that the divergence-less condition may be enforced by claiming that $\mathbf{K}$ may be written in the form

$$\mathbf{K} = -\hat{z} \times \nabla\sigma \qquad (10.2)$$

with σ being a scalar function (the stream function) evaluated on the super-conductor. The reader may easily check[2] that

   (i) $\nabla \cdot \mathbf{K} = 0$ is automatically satisfied.
   (ii) By imposing $\sigma = $ constant on the boundaries, one obtains that $\mathbf{K}$ flows along them as required.
   (iii) The contour lines σ give the streamlines of $\mathbf{K}$, or equivalently: $\mathbf{K} \cdot \nabla\sigma = 0$.

Then, the physical model is built by posing the Maxwell equations together with the superconducting material law replacing the current density by the expression

---

[2] See exercise R10-1.

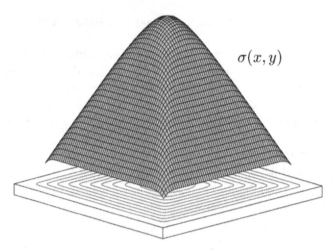

$\sigma(x, y)$

Figure 10.2. The stream function above the platelet of figure 10.1. For visual purposes, the lower part of $\sigma(x, y)$ is removed. 🔗. Source codes available at https://iopscience.iop.org/book/978-0-7503-2711-4.

$-\hat{z} \times \nabla \sigma$. One ends up with a formulation in terms of the function $\sigma(x, y)$. Eventually, backsubstitution allows to obtain all the physical quantities of interest in terms of $\sigma(x, y)$.

Just for illustration, figure 10.2 shows the stream function related to the current density streamlines in figure 10.1. Notice that, as said, the contour levels of σ correspond to the loops of current circulating on the plate. Notice also that, arbitrarily, the choice $\sigma(x, y) = 0$ along the sample's boundary was made.

10.1.2 The critical state in thin samples

In this section, we will forward the explicit formulation of the critical state problem in superconducting platelets. We apply the variational statement introduced in chapter 8. To start with, we recall that, in terms of σ, equation (8.1) becomes

$$
\mathcal{F}[\sigma_{n+1}] \equiv \frac{1}{2} \iint_\Omega \left[\frac{\nabla \sigma_{n+1}(\mathbf{r}) \cdot \nabla \sigma_{n+1}(\mathbf{r}')}{\|\mathbf{r} - \mathbf{r}'\|} - 2 \frac{\nabla \sigma_n(\mathbf{r}) \cdot \nabla \sigma_{n+1}(\mathbf{r}')}{\|\mathbf{r} - \mathbf{r}'\|} \right] d\mathbf{r}\, d\mathbf{r}'
$$
$$
+ 4\pi (H_z^{a,n+1} - H_z^{a,n}) \int_\Omega \sigma_{n+1}(\mathbf{r})\, d\mathbf{r} \tag{10.3}
$$

with $H_z^{a,n}$ being the z-component of the uniform applied external field, and the subindices n and $n + 1$ used to indicate two successive time layers.

The minimisation of $\mathcal{F}[\sigma]$ under the critical state restriction[3] $K \leqslant K_c$ will be performed numerically. Below, we present the discretised version of equation (10.3) and provide the scripts for obtaining its solution.

---

[3] In general, the superconducting material law must be stated in a more general manner. For instance, if the main mechanism is flux depinning one has to impose $K_\perp \leqslant K_c$ with $K_\perp$ being the component of $\mathbf{K}$ perpendicular to the magnetic field, and only for 'vertical' external fields one may identify $K = K_\perp$. This will be discussed in section 10.2.

10.1.3 Stream functions: discrete formulation

To start with, we define a grid of points that will be the domain of our discretised statement (i.e. the superconducting surface). As shown in figure 10.3, for the case of a square platelet, a convenient selection may be a uniform square mesh of points in the XY plane. Meshing techniques for general geometries are an important and rather specialised topic beyond the scope of this book, that should be addressed in case of dealing with other geometries[4].

In our notation, the domain of the function $\sigma(x, y)$ is subdivided into four quarters of grid points. Each set will be collected in an array indexed by $i = 1, 2, \ldots, N \times N$. Thus, we define the $4N^2$-vector $|\sigma\rangle$, whose components are nothing but the values of the function at the points of the grid, i.e. $\sigma_i = \sigma(x_i, y_i)$. Under the basic assumption that one is using a sufficiently refined mesh, the discrete minimisation principle arising from (10.3) is a straightforward generalisation of (8.26) given by

$$\text{minimise} \quad \mathcal{F}(\sigma) = \frac{1}{2}\langle\sigma|\mathbf{Q}|\sigma\rangle - \langle\sigma^\vee|\mathbf{Q}|\sigma\rangle + 4\pi\langle\Delta H_z^a|\sigma\rangle \tag{10.4}$$

with the basic replacement of the mutual inductance matrix $\mathbf{m}$ by the new matrix $\mathbf{Q}$ and ΔH_z^a the increment of perpendicular magnetic field. As the reader may check, the structure of our new 'coupling matrix' is as follows

$$\mathbf{Q} = \mathbf{D}_x{}^\mathrm{T}\,\mathbf{m}_x\,\mathbf{D}_x + \mathbf{D}_y{}^\mathrm{T}\,\mathbf{m}_y\,\mathbf{D}_y \tag{10.5}$$

Here, $\mathbf{D}_x$, $\mathbf{D}_y$ are two matrices whose elements take the values ± 1, 0 (normalising units by the grid spacing) and provide the discrete values of the sheet current, i.e.

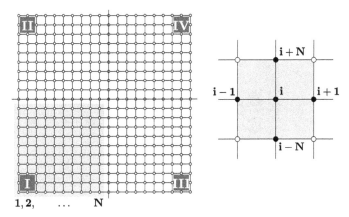

Figure 10.3. Numerical evaluations on the grid. As illustrated, points are labelled by indices in every quarter of the platelet.

---

[4] A thorough treatment of the general problem may be found in the book by A Bossavit [7].

$$|\mathbf{K}_x\rangle \rightarrow \frac{\partial|\sigma\rangle}{\partial y} \equiv \mathbf{D}_x|\sigma\rangle; \quad |\mathbf{K}_y\rangle \rightarrow -\frac{\partial|\sigma\rangle}{\partial x} \equiv \mathbf{D}_y|\sigma\rangle \qquad (10.6)$$

$\mathbf{D}_x^T$, $\mathbf{D}_y^T$ are their transposes, and $\mathbf{m}_x$, $\mathbf{m}_y$ stand for the mutual inductance matrices coupling either horizontal or vertical elements of current at the points of the grid[5].

Resorting to figures 10.3 and 10.4 one may show that the nonzero elements of the *finite difference* $\mathbf{D}$ matrices are given by[6]

$$\begin{aligned}
\mathbf{D}_x(i, i) &= 1 \quad ; \quad \mathbf{D}_x(i, i + N) = -1 \\
\mathbf{D}_y(i, i) &= -1 \quad ; \quad \mathbf{D}_y(i, i + 1) = 1
\end{aligned} \qquad (10.7)$$

whereas the mutual inductance elements are given by

$$\begin{aligned}
\mathbf{m}_x(i, j) &= \mathbf{m}_y(i, j) \approx L_i L_j / d_{ij}, \quad i \neq j \\
\mathbf{m}_x(i, i) &= \mathbf{m}_y(i, i) \approx 2 \left\{ L_i \log \left[\zeta_i \left(1 + \sqrt{1 + 1/\zeta_i^2} \right) \right] \right\}.
\end{aligned} \qquad (10.8)$$

as obtained from Neumann's formula: $\mathbf{M}_{ij} = (\mu_0/4\pi) \iint d\ell_i \cdot d\ell_j / \|\mathbf{r}_i - \mathbf{r}_j\|$ and using the self-inductance of a segment of current (see the classical reference [8] and the book by Grover [9]). L_i stands for the length of the segment, and ζ_i for its aspect ratio (length/width). For simplicity, we will use $\zeta_i = 1$, which means that the aspect ratio of our platelet is given by $a/d = N$. So that, the higher the value of N, the better the 2D approximation.

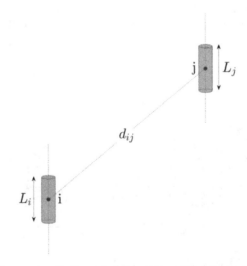

Figure 10.4. Definition of parameters for the mutual inductance matrix between elementary current segments (see text).

The actual application of equation (10.4) still needs to clarify some details. To start with, one must recall the constraint on the critical current density. In the matrix notation introduced above, and according to equation (10.6), the critical state restriction reads

$$|\mathbf{K}_x\rangle_i^2 + |\mathbf{K}_y\rangle_i^2 = \mathbf{D}_x|\sigma\rangle_i^2 + \mathbf{D}_y|\sigma\rangle_i^2 \leqslant K_{ci}^2, \quad \forall i \tag{10.9}$$

which means that one may impose a different restriction for the current density at each point of the grid through a non-uniform set of values K_{ci}.

Eventually, a key element in the application of equations (10.4) and (10.9) is that, taking advantage of symmetry, one may noticeably reduce the computational effort. In particular, it is apparent that one may solve the problem just by considering one quarter of the full number of variables. Thus, by setting the origin of coordinates at the centre of the platelet, it is obvious that one has

$$\begin{aligned} K_x(x, y) &= -K_x(x, -y) = K_x(-x, y) = -K_x(-x - y) \\ K_y(x, y) &= K_y(x, -y) = -K_y(-x, y) = -K_y(-x, -y) \end{aligned} \tag{10.10}$$

or, equivalently (as it is apparent in figure 10.2)

$$\sigma(x, y) = \sigma(x, -y) = \sigma(-x, y) = \sigma(-x - y) \tag{10.11}$$

This means that one may solve the problem with the unknowns of region I (for instance) and then 'replicate' the values for the points of regions II, III and IV (see figure 10.3).

In practice, the implementation of symmetry without losing any information about the inductive coupling between all the elements of the mesh may be done as follows. One solves the problem for σ at the points $i = 1, 2, \ldots, N \times N$ in the first quarter, by using:

$$\begin{aligned} \mathbf{m}_x &= 2\mathbf{m}_x^{11} + \mathbf{m}_x^{13} + \mathbf{m}_x^{24} - \mathbf{m}_x^{12} - \mathbf{m}_x^{14} - \mathbf{m}_x^{34} - \mathbf{m}_x^{24} \\ \mathbf{m}_y &= 2\mathbf{m}_y^{11} + \mathbf{m}_y^{12} + \mathbf{m}_y^{34} - \mathbf{m}_y^{14} - \mathbf{m}_y^{13} - \mathbf{m}_y^{24} - \mathbf{m}_y^{23} \end{aligned} \tag{10.12}$$

In this equation, superindices are used to indicate the sector implied. For instance, $\mathbf{m}_x^{14}$ means that the elements $\mathbf{m}_x(i_1, j_4)$ with (x_{i_1}, y_{i_1}) a point in I and (x_{j_4}, y_{j_4}) a point in IV are considered.

Summarising this section, the evolutionary critical state flux penetration profiles in a planar sample under a dynamic external magnetic field may be obtained through a variational principle (10.4) for the so-called stream function σ. With $\sigma(x, y)$ at hand, one may evaluate the electromagnetic quantities of interest. Actual examples of this and the related software are given in the next section.

10.2 Response to applied magnetic fields

Our interest here concerns the application of magnetic fields either fully perpendicular or at least partially perpendicular (say, inclined) to the surface very thin samples. The case of parallel magnetic fields has already been described in section 3.3 as the trivial application of Bean's model.

To start with we use the stream function formalism so as to determine the superconducting currents that arise when a given magnetic field process takes place. As some fundamental differences arise in the treatment of either perpendicular or inclined fields they will be presented in two different subsections. Noticeably, with $\sigma(x, y)$ at hand, one may analyse a number of related physical properties. In particular, one may predict two celebrated experimental quantities: the sample's magnetic moment and the flux density pattern on its surface.

10.2.1 Perpendicular fields

Figure 10.5 displays the current density streamlines obtained as contour plots of the stream function $\sigma(x, y)$ that solves equation (10.4) for the external field process

$$H_x^a = H_y^a = 0; \quad H_z^a: 0 \to H_m^a \to -H_m^a \to H_m^a, \tag{10.13}$$

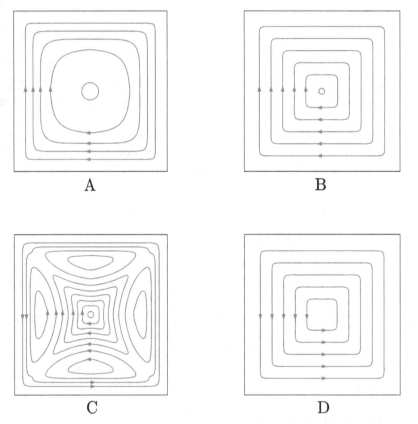

Figure 10.5. Current density streamlines induced by an applied field perpendicular to the sample (conditions sketched in figure 10.1). The field is cycled between $H_z^a = 5H_{cs}$ and $H_z^a = -5H_{cs}$ (see text). A and B correspond to the increasing branch ($H_z^a/H_{cs} = 1.25$, 5.0) respectively, while C and D correspond to the decreasing branch ($H_z^a/H_{cs} = 3.25$, −5.0). $\mathcal{O}$. Source codes available at https://iopscience.iop.org/book/978-0-7503-2711-4.

i.e. a perpendicular flux density is applied with the z-component of the magnetic field cycled between the extreme values $\pm H_m^a$. In order to reach the full penetration regime, we have used $H_m^a = 5H_{cs}$, with $H_{cs} = K_c/\pi$.[7]

As detailed in the caption, the profiles A and B correspond to the initial magnetisation process $H_{A,B} \in [0 \rightarrow H_m^a]$, and C and D to the descending branch $H_{C,D} \in [H_m^a \rightarrow -H_m^a]$. Several aspects deserve to be mentioned:

(i) As expected, superconducting currents circulate in the form of loops that screen the variations of the applied magnetic field and reach a saturation profile when the applied field increases enough (panel B). This saturation corresponds to the condition $K = K_c$ everywhere in the sample and gives way to square loops of current. When the excitation is reversed, an outer layer with counter-circulating currents appears (panel C), and eventually fills the sample.

(ii) As the reader may show (see exercise R10-2), the distance between two nearby lines of current density relates to the value of K in the vicinity. According to this, at some stages of the process (see panel A) the critical state ansatz does not seem fulfilled; one may identify regions with $K < K_c$. However, this is not a flaw of the calculation! The physical picture of this result is as follows. One has to recall the underlying 3D nature of the real problem. In fact, truly, one has $J = \pm J_c$, 0, and $K < K_c$ is nothing but the result of an average along a z-line for which one has $J = J_c$ at some points and $J = 0$ at others. In order to further clarify this aspect, we suggest to review exercise C9-1 and to work out the topic in R10-3.

The scripts used to generate figure 10.5 will be presented in the forthcoming section in conjunction with those of figure 10.9. At this point we proceed by analysing the macroscopic counterpart of the above results, i.e. the implication of the simulated current distributions in some straightforward experimental quantities.

10.2.2 Magnetic hysteresis

As said before, a number of physical quantities may be straightforwardly related to the stream function $\sigma(x, y)$. Thus, as proposed in exercise *R10-4*, one may show that the sample's magnetic moment is given by

$$m_z = \int_\Omega \sigma(x, y)\, dxdy, \tag{10.14}$$

with Ω the surface of the planar sample, i.e. σ is equivalent to a density of magnetic moment. In dimensionless units[8], the following expression gives the approximation to the sample's volume magnetisation in terms of the discrete values of the stream function

---

[7] For comparison to more 'conventional' units, consider that $K_c = J_c d$, which establishes the relation to the 'Bean-like' penetration field $H_p = J_c a/2$. Thus: $H_{cs} = (2d/\pi a)H_p$.
[8] We normalise distances by d, K by K_c, H by $H_c = K_c/\pi$ and σ by $K_c d$.

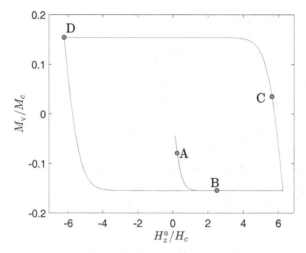

Figure 10.6. Hysteresis cycle for the square platelet in perpendicular field. We have used normalised units for the volume magnetisation and applied magnetic field, through the definitions $H_c \equiv J_c d / \pi$ and $M_c \equiv J_c a$. ⌕. Source codes available at https://iopscience.iop.org/book/978-0-7503-2711-4.

$$\frac{M_v}{J_c a} = \left(\frac{d}{a}\right)^3 \sum_i \sigma(i) \qquad (10.15)$$

The magnetisation loop that corresponds to the applied field cycle in equation (10.13) is shown in figure 10.6. On the curve, we have marked the points corresponding to the specific values selected in figure 10.5.

10.2.3 On-surface magnetic flux density

In section 4.3 we recalled that magneto-optics has been a highly successful technique for the investigation of thin superconductors. As said, the experiments may reveal with extreme resolution how the magnetic flux density distributes above the sample (which may give rich physical information). Although the quantitative aspects of such experiments will be analysed in chapter 11 with some detail, here we advance some qualitative results of interest. To start with, we notice that relying on the knowledge of $\sigma(x, y)$ one may straightforwardly evaluate the normal component of the magnetic flux density B_z above the sample[9], which is in fact the quantity that reveals in conventional magneto-optics. Thus, starting with $\mathbf{B} = \nabla \times \mathbf{A}$, and by using the linear relation between the vector potential and the currents, one may show that, numerically, B_z may be obtained on the sample's surface as follows:

$$|\mathbf{B}_z\rangle_0 = \mathbf{Z}_0 |\sigma\rangle + \mu_0 | \mathbf{H}_z^a\rangle_0 \qquad (10.16)$$

with $\mathbf{Z}_0$ being the *geometrical* matrix

---

[9] Here, we will only display $B_z(x, y, 0)$, i.e. the normal component, just on surface, but as suggested in exercise C10-2, the method may be generalised to obtain $\mathbf{B}(x, y, z)$.

$$\mathbf{Z}_0 = -(\mathbf{D}_x \, \mathbf{m}_x \, \mathbf{D}_x + \mathbf{D}_y \, \mathbf{m}_y \, \mathbf{D}_y) \tag{10.17}$$

$|\mathbf{B}_z\rangle_0$ standing for the collection of values of the full magnetic field $B_z(x_i, y_i)$ on surface, and $|\mathbf{H}_z^a\rangle_0$ for the applied magnetic field at such points of the superconductor. We recall that this method is similar to the technique used by Grant and co-workers in [10], based on an equivalent mesh of local dipoles for the current.

Figure 10.7 shows the flux density profiles corresponding to the magnetisation process described by the relation (10.13) in the stages A, B, C, D. For comparison with experimental magnetooptical images one may refer to the book by Johansen and Shantsev [11]. As described in experiments, notice that magnetic flux penetrates predominantly along the central part of the edges, producing the star-like (sometimes called cushion-like) patterns. It is of mention that, related to the quasi-2D nature of the problem, the current density streamlines noticeably differ from the flux penetration fronts[10].

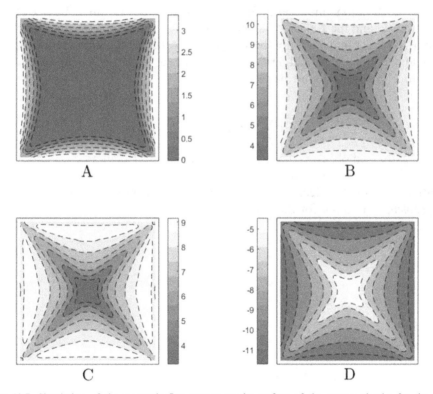

Figure 10.7. Simulation of the magnetic flux pattern at the surface of the square platelet for the same experiment considered in figure 10.5. The cushion-like pattern of the penetration fronts is emphasised. 🔗. Source codes available at https://iopscience.iop.org/book/978-0-7503-2711-4.

---

[10] Recall that in the opposite limit, i.e. long bar for which $a \ll d$, both current and magnetic flux penetrate with identical fronts that reproduce the shape of the sample's boundary.

10.2.4 Application of inclined magnetic fields

Experiments involving magnetic fields applied with some inclination relative to the surface of the sample have been very instructive in the investigation of vortex matter in high-T_c materials (see for instance, the experimental observations by Vlasko-Vlasov in [12]). In particular, as said before, a fundamental physical feature introduced by such configurations is that $\mathbf{B}$ and $\mathbf{K}$ are not perpendicular. Thus, by using inclined fields, one may probe the relevance of dissipation phenomena other than those related to flux depinning (caused by the Lorentz-like force $\mathbf{J} \times \mathbf{B}$). Here, we limit ourselves to the situation in which depinning is predominant and analyse its consequences in the predicted phenomenology. Figure 10.8 displays a sketch of the configuration considered. Under the mentioned assumption the dynamics of the critical state will be obtained by solving equation (10.4) with the constraint:

$$K_\perp^2 \leqslant K_c^2 \Rightarrow K_x^2 + K_y^2 \cos^2 \theta \leqslant K_c^2 \tag{10.18}$$

It is important to realise that in the above equation θ is the angle between the full magnetic field and the z-axis. If the self field of the superconductor is not negligible, one must consider that θ it is not the straightforward experimental parameter defined by the applied magnetic field $\mu_0 \mathbf{H}^a$ and the z-axis. In fact, if one denotes by θ_0 such angle, equation (10.18) must be replaced by

$$K_x^2 + K_y^2 \left(1 - \frac{B_0^2 \sin^2 \theta_0}{B_0^2 + 2 B_0 B_{\text{self}} \cos \theta_0 + B_{\text{self}}^2} \right) \leqslant K_c^2 \tag{10.19}$$

with B_{self} the sample's magnetic field, that one may calculate by using (10.16). Nevertheless, the incorporation of this fact to the theory requires to pay some attention, because the effect should be introduced self-consistently, as the solution of the problem is an input for the restriction!

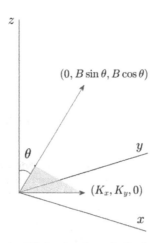

Figure 10.8. Superconducting platelet (at XY plane) under an inclined magnetic field. The components of the surface current density parallel and perpendicular to $\mathbf{B}$ are emphasised.

In general, when the applied magnetic field intensity $\mathbf{H}^a$ is much above $H_c \equiv J_c d$ one may safely use equation (10.18). Physically, H_c parameterises the 'saturated' critical state of the sample. For such situation, one may even find semi-analytical treatments in the literature (as in the extensive work by Brandt and Mikitik [6]). Here, aiming at a first approximation to the full problem, we will provide the general numerical tools for dealing with inclined fields, but restrict our attention to the case $H^a \gg H_c$. Figure 10.9 displays the induced current density streamlines and the corresponding flux density profiles for two angles of the process:

$$\text{1st)} \quad H_x^a = H_y^a = 0; \; H_z^a: 0 \rightarrow H_m^a = \frac{125}{\pi} H_c$$

$$\text{2nd)} \quad H_x^a = 0; \; H_y^a: 0 \rightarrow H_m^a \sin \theta; \; H_z^a: H_m^a \rightarrow H_m^a \cos \theta \tag{10.20}$$

It is noticeable that:
- (i) Already for a small angle ($\theta = \pi/40$) one may observe the formation of loops of current between the inner trapped field and the peripheral counter-circulating region.
- (ii) The inclination of the applied field towards the y-axis produces a clear anisotropy in the response of the sample. As expected, K_x flows with

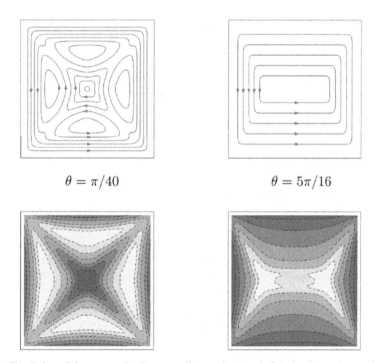

$$\theta = \pi/40 \qquad\qquad\qquad \theta = 5\pi/16$$

Figure 10.9. Simulation of the current density streamlines and magnetic flux density pattern at the surface of the square platelet for the experiment sketched in figure 10.8 and described in (10.20). $\wp$. Source codes available at https://iopscience.iop.org/book/978-0-7503-2711-4.

smaller magnitude because the magnetic field directs towards the perpendicular direction and we are using a model that only restricts $K_\perp$.

(iii) The induced anisotropy is evident in the simulated flux density plots. As named in the literature, the cushion-like structure now becomes a *bow-tie*.

10.3 Description of the numerical resources

In this section we describe the scripts that have been used to produce the figures of this chapter. Specifically, the resources behind figures 10.5, 10.7 and 10.9, whose nucleus is the application of equation (10.4) will be explained with some detail. They will be the basis for the rest of figures, and possibly for further studies. Calculations have been performed by means of the following list of scripts

- h_perp_plate.m, h_incl_plate.m (main scripts for figures 10.5 and 10.9)
- geom_plate.m
- matrixM_sym_cuad.m, self_M.m, mutual_M.m
- Q_from_M.m
- Kritical.m, Kritical_inclined.m
- cost.m
- figure_10_5.m
- figure_10_6.m, magnetization.m
- figure_10_7.m, Bz_profile.m
- figure_10_9.m, Bz_profile_inclined.m

The main program creates the specific problem through a number of auxiliary scripts and functions, according to the structure detailed below.

Geometry and geometrical operators: geom_plate.m defines the arrays containing the (x_i, y_i) coordinates of the points that discretise the problem (see figure 10.3). We stress that interior and boundary points of the grid are to be distinguished. matrixM_sym_cuad.m produces the components of the mutual inductance matrices $\mathbf{m}_x, \mathbf{m}_y$ in equation (10.12) based on the expressions in (10.8). Q_from_M.m gives the components of the matrices $\mathbf{D}_x$, $\mathbf{D}_y$ and $\mathbf{Q}$ in equation (10.5), that will act as the discrete version of the corresponding differential operators.

Minimisation: based on internal MATLAB functions h_perp_plate.m and h_incl_plate.m apply equation (10.4) through the minimisation of the function cost.m either constrained by relation (10.9), i.e. Kritical.m or (10.19), ie.: Kritical_inclined.m, respectively.

Post-processing: with the $\sigma(x, y)$ data at hand (in the form of an array of values $|\sigma\rangle$ at the points (x_i, y_i)), one may produce a number of plots as those in figures 10.2, 10.5, 10.6, 10.7, and 10.9. Basically, post-processing is performed by application of equations (10.15) and (10.16) and taking advantage of MATLAB built-in functions as griddata(), contour() and smooth(), and the dedicated scripts in Bz_profile.m and Bz_profile_inclined.m. It must be mentioned that smoothing is just an option for visual purposes which should be handled with care so as to avoid 'artificially' rounded curves. Incidentally, we comment that the arrow heads indicating the flow of current in figures 10.5 and 10.9 have been created from the scratch as one may find in the corresponding scripts. They are built upon the ideas introduced in section 6.1.1.

10.4 Review exercises and challenges

10.4.1 Review exercises

Review exercise R10-1

Verify the properties (i)–(iii) following from equation (10.2) that introduces the stream function $\sigma(x, y)$.

Hint: recall the geometric properties of the gradient of a scalar function, in particular, that the contour lines of a scalar function are perpendicular to its gradient.

Review exercise R10-2

Show that the distance between the contour levels of the stream function is proportional to the modulus of the sheet current in the vicinity of a given point.

Hint: consider that in general $d\sigma = \nabla\sigma \cdot d\ell$, that the gradient of a function is perpendicular to its isolines, and that $K^2 = \|\nabla\sigma\|^2$.

Review exercise R10-3

By using the configuration introduced in section 8.3 show that the thin sample limit developed in the current chapter may be conceptualised as the approximation to the 3D problem, for which the sheet current **K** is physically significant. For this, solve the exact critical state problem for a cylinder with different values of L/R, calculate $K(r) = \int J(r, z)dz$ and plot K *versus* r together with the penetration profile as in figure 8.5.

Hint: integration may be performed numerically by a simple summation of values of the current density array, separated by N positions (see figure 8.4). Reviewing exercise C9-1 may also be enlightening.

Review exercise R10-4

Derive the relation between the magnetic moment of the superconducting platelet m_z and the related stream function $\sigma(x, y)$, i.e. prove relation (10.14).

Hint: recall the integral expression for the sample's magnetic moment (equation (8.28)), consider the average across the thickness and use the definition of σ.

Review exercise R10-5

Show that for the case of a square platelet (side a) in the *saturated* critical state under perpendicular field, the following relation holds for the volume magnetisation

$$M_V = J_c \, a/6 \qquad\qquad (10.21)$$

Hint: start with equation (10.14) and consider the expression for the volume of a pyramid with square basis.

10.4.2 Challenges

Exercise C10-1

By properly changing the variational principle in equation (10.4), obtain the formulation of the Meissner state for a square platelet. Reproduce figures 10.5 and 10.7 in such case.

Hint: replace the critical state minimisation functional $\mathcal{F}(\sigma)$ by the corresponding $U(\sigma)$ for the Meissner state (see equations (13.16) and (13.17)).

Exercise C10-2

Show that equation (10.16) may be generalised to

$$|B_x\rangle = X|\sigma\rangle + \mu_0|H_x\rangle_{\text{Applied}}$$
$$|B_y\rangle = Y|\sigma\rangle + \mu_0|H_x\rangle_{\text{Applied}} \qquad (10.22)$$
$$|B_z\rangle = Z|\sigma\rangle + \mu_0|H_z\rangle_{\text{Applied}}$$

so as to obtain the magnetic field at any point of space, with X, Y, Z geometrical matrices that extend Z_0.[11]

Hint: consider the relation $B = \nabla \times A$, equations (8.50) and (10.2), and the definition of the D matrices.

---

[11] These matrices will be named after 'Biot-Savart' operators, as they act on the current density and give way to the magnetic field.

Exercise C10-3

Investigate the influence of self-field effects in the response of the superconducting platelet to inclined magnetic fields.

Hint: one must use relation (10.19) instead of (10.18). Do this by modifying the scripts `h_incl_plate.m` and `Kritical_inclined.m`. As the constraint depends on the solution itself, implement some consistency procedure. For example, use small steps in the applied field and estimate B_{self} from the previous step.

References

[1] Mikheenko P N and Kuzovlev Y E 1993 Inductance measurements of HTSC films with high critical currents *Physica C: Supercond. Appl.* **204** 229–36

[2] Pearl J 1964 Current distribution in superconducting films carrying quantized fluxoids *Appl. Phys. Lett.* **5** 65–6

[3] Prigozhin L 1998 Solution of thin film magnetization problems in type-II superconductivity *J. Comput. Phys.* **144** 180–93

[4] Brandt E H 1995 Square and rectangular thin superconductors in a transverse magnetic field *Phys. Rev. Lett.* **74** 3025–8

[5] Brandt E H and Mikitik G P 2007 Vortex shaking in superconducting platelets in an inclined magnetic field *Supercond. Sci. Technol.* **20** S111–6

[6] Mikitik G P 2010 Critical states in thin planar type-II superconductors in a perpendicular or inclined magnetic field (review) *Low Temp. Phys.* **36** 13–38

[7] Bossavit A 1998 *Computational Electromagnetism: Variational Formulations, Complementarity, Edge Elements* (New York: Academic)

[8] Rosa E B 1907 The self and mutual inductances of linear conductors *Bull. Bur. Stand.* **4** 301

[9] Grover F W 1980 *Inductance calculations* (Research Triangle Park, NC: Instrument Society of America)

[10] Grant P D, Denhoff M W, Xing W, Brown P, Govorkov S, Irwin J C, Heinrich B, Zhou H, Fife A A and Cragg A R 1994 Determination of current and flux distribution in squares of thin-film high-temperature superconductors *Physica C: Supercond. Appl.* **229** 289–300

[11] Johansen T H and Shantsev D V 2004 *Magneto-Optical Imaging. II. Mathematics, Physics and Chemistry* vol 142 Nato Science Series 01 (Dordrecht: Springer)

[12] Vlasko-Vlasov V K, Glatz A, Koshelev A E, Welp U and Kwok W K 2015 Anisotropic superconductors in tilted magnetic fields *Phys. Rev.* B **91** 224505

IOP Publishing

Macroscopic Superconducting Phenomena
An interactive guide
Antonio Badía-Majós

Chapter 11

Magneto-optical imaging of superconductors

As emphasised by Bending [1], local magnetic probes have been successfully used to extract important superconducting parameters in recent years. Among them, magneto-optical imaging[1] has allowed to elucidate many aspects related to the spatial distribution of supercurrents, the determination of microscopic parameters as the first penetration field, and even to visualise individual vortices and their dynamics [2, 3]. All this has been enabled by the progress in MOI techniques, as well as by the noticeable issue of analytical and numerical methods for the interpretation of experiments (recall that the fundamentals of the technique may be found in section 4.3).

In this chapter we introduce some tools for the simulation and study of MOI experiments, so as to provide a basic rationale on the topic. Though dealing with a local probe technique, we still keep our analysis at the macroscopic level, and focus on such experiments with a resolution that averages over many vortices and treats the local current density as a smooth function. A number of analytical and numerical tools (all based on MATLAB scripts) are introduced that allow to deal with the customary sample geometry, i.e. long beams and rectangular platelets exposed to uniform fields. The Meissner state and critical state regimes will be considered. Throughout the chapter, the main physical quantity is the magnetic flux across the surface of the sample.

11.1 Magneto-optics in the Meissner state

In this section we will introduce a method for studying the magneto-optical response of superconducting beams in the Meissner state. The experimental geometry is shown in figure 11.1. A garnet indicator is placed on top of a long superconducting beam subject to an applied magnetic field perpendicular to its surface. As said in section 4.3, MO images will reveal the flux density profiles perpendicular to the

---

[1] We will use MOI for magneto-optical imaging throughout the chapter.

doi:10.1088/978-0-7503-2711-4ch11

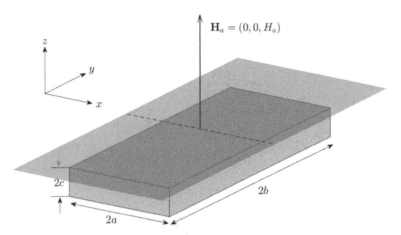

Figure 11.1. Sketch of the experimental arrangement for the conventional MOI experiment. A sample in the form of a beam ($-a < x < a$, $-b < y < b$, $-c < z < c$) is exposed to a uniform magnetic field $\mu_0 \mathbf{H}_a$ along the z-axis. Laying above the sample is a ferrimagnetic garnet indicator that reveals the structure $B_z(x, y, z_0)$.

garnet, i.e. $B_z(z = h + c/2)$ with h being the height of the indicator over the superconductor.

Here, taking advantage of the long sample limit[2] we will introduce the analytical formulation by Brandt and Mikitik [4] concerning the Meissner state for super-conductors with rectangular cross section. On the other hand, relying on the numerical methods introduced in chapter 9, a numerical solution for finite geometries may also be obtained (guided exercise R11-2).

11.1.1 Physical modelling

Following [4], the perfect Meissner state of the long superconducting beam under uniform perpendicular field takes place with a translationally symmetric solution with a surface current density that may be written in the form $(0, -K(x, y), 0)$ and is given by

$$K(x, z = \pm c) = \frac{H_a S_x(x)}{\sqrt{1 - S_x(x)^2}}$$

$$K(x = \pm a, z) = \frac{\pm H_a \sqrt{1 - m S_z(z)^2}}{\sqrt{m(1 - S_z(z)^2)}} \qquad (11.1)$$

The plots of such dependencies are shown in figures 11.2 and 11.3.

Above, the following definitions have been used:

- m is a geometry dependent parameter that depends on the sample's aspect ratio $\zeta = c/a$. It is obtained by solving the equation

---

[2] Theoretically, we use the ansatz $b \gg a$, c, that imposes translational symmetry along the y-axis. In practice, it is known that the behaviour close to the midplane $y = 0$ of a finite sample is basically independent of y as long as $b > 2a$.

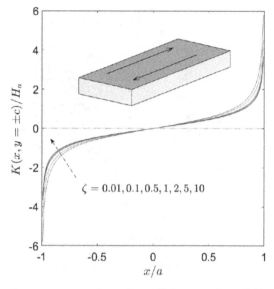

Figure 11.2. Meissner surface currents: upper/lower faces. 🔗. Source codes available at https://iopscience.iop.org/book/978-0-7503-2711-4.

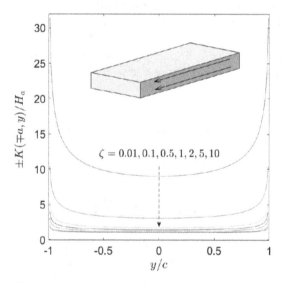

Figure 11.3. Meissner surface currents: lateral faces. 🔗. Source codes available at https://iopscience.iop.org/book/978-0-7503-2711-4.

$$f(1, m) = \zeta f(1, 1 - m) \qquad (11.2)$$

with $f(s, m) = E[\sin^{-1}(s), \sqrt{m}] - (1 - m)F[\sin^{-1}(s), \sqrt{m}]$ and E and F being incomplete elliptic integrals.
- With m at hand, $S_x(x)$ and $S_z(z)$ are determined by solving

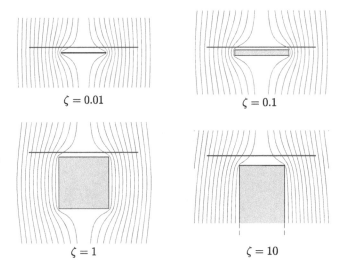

$\zeta = 0.01$ $\zeta = 0.1$

$\zeta = 1$ $\zeta = 10$

Figure 11.4. Magnetic field lines surrounding ideal superconducting beams in the Meissner state. Long dimension is across the paper. ζ indicates the aspect ratio of their rectangular cross section. The lines above the superconductor are indicative of the position of the garnet in the MOI experiments. ⬚. Source codes available at https://iopscience.iop.org/book/978-0-7503-2711-4.

$$f(S_x, 1 - m) = \frac{x}{a} f(1, 1 - m)$$

$$f(S_z, m) = \frac{z}{c} f(1, m)$$

(11.3)

Dedicated MATLAB codes for solving equations (11.2) and (11.3) are provided.

Eventually, the magnetic field around the sample may be calculated from Ampère's law by numerical integration with the above dependencies of the functions S_x, S_y determining the current density. Thus, figure 11.4 shows the results for different values of the aspect ratio ζ. For further consideration, we indicate the typical position of the garnets that will reveal the value of B_z at the given height.

11.1.2 Quantification of the MOI signal

As was discussed in [5], the quantitative analysis of the MOI signal, i.e. the actual dependence of $B_z(x, y = 0, z = h + c/2)$ is strongly influenced by the actual value of the observation distance h. In fact, this is a critical parameter whose uncertainty introduces difficulties in the determination of the superconducting properties. For instance, as concerns the penetration fields, one should consider that the flux profile above a certain sample depends on the applied field H_a as well as on the value of the height h, which is very important at small distances. Thus, below, we provide some tools that allow to analyse the problem for a given geometry. Figure 11.5 shows the distribution of B_z (in fact, of the superconducting contribution B_z^{sc}) for different heights above the sample. We plot the flux density that would be recorded by scanning along the central line (Figure 11.1) for a sample with aspect ratio $\zeta = 0.1$.

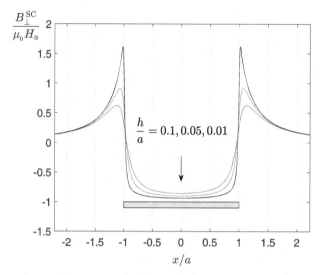

Figure 11.5. Distribution of $B_\perp^{sc}$ across the width of the superconducting beam of aspect ratio $\zeta = 0.1$ at three different heights above the surface: $h/a = 0.1$ (red), $h/a = 0.05$ (blue), and $h/a = 0.01$ (black). 🔗. Source codes available at https://iopscience.iop.org/book/978-0-7503-2711-4.

As the reader may notice, the closer to the sample, the sharper the peaks. Also, one may notice that the position of the peaks becomes closer to the sample edges as the observation height diminishes.

Figures 11.2, 11.3, 11.4, 11.5 have been created with the following scripts

- `get_m.m, inverter_brandt.m` (obtain m)
- `get_Sx_Sz.m, inverter_x.m, inverter_z.m` (obtain S_x, S_z)
- `figure_11_2.m`
- `figure_11_3.m`
- `read_files.m`
- `figure_11_4.m` (plot contour lines of vector potential)
- `integrate_field.m` (apply Ampère's law to the filamentary currents and integrate within faces, i.e. Equation (11.6))
- `figure_11_5.m` (plot $B_z(x, y = 0, z = h + c/2)$).

11.2 Magneto-optics in the critical state

When the magnetic flux penetrates the superconductor (either partially or totally) the $B_\perp$ profiles at a certain observation distance reveal a rather different structure to that of figure 11.5. In order to illustrate this question and allow further investigation, this section introduces some resources for simulating MOI in the critical state. For more specialised information on this topic we recommend to read the extensive work by Jooss and collaborators [3]. Here, we will concentrate on two geometries that allow a relatively simple treatment, i.e. long strips and rectangular platelets under perpendicular field. In fact, real experiments are mostly realised with flat samples of finite sizes.

11.2.1 Thin sample limit: long strips

Let us consider the geometry depicted in figure 11.1 in the limiting case $b \gg a \gg c$. The problem is quasi-1D and the superconducting response may be described in terms of the sheet current[3] along y-axis, which we write in the form $(0, -K(x), 0)$. As shown by Brandt and Indenbom [6], in such approximation, the critical state problem admits the analytical solution[4]

$$
K(x) = \begin{cases} \dfrac{2K_c}{\pi} \tan^{-1}\left[\dfrac{c_{B}x}{\sqrt{b_{B}^{2} - x^{2}}}\right]; & |x| < b_{B} \\[4mm] K_c \dfrac{x}{|x|} & ; \quad b_{B} \leqslant |x| \leqslant a \end{cases}
\tag{11.4}
$$

for the initial magnetisation of the strip. Here we have used the definitions

$$
\begin{aligned}
K_c &\equiv 2cJ_c; \quad H_c \equiv K_c/\pi \\
b_{B} &\equiv a/\cosh(H_a/K_c) \\
c_{B} &\equiv \tanh(H_a/H_c)
\end{aligned}
\tag{11.5}
$$

Integration of Ampère's law for elementary wires that compose the above sheet current straightforwardly leads to obtaining the flux profiles shown in figure 11.6. This entails to apply:

$$
B_z(x', z') = \frac{\mu_0}{2\pi} \int_{-a}^{a} \frac{x - x'}{z'^2 + (x - x')^2} K_y(x)dx
\tag{11.6}
$$

We recall the change of curvature for the inner part of the peaks related to the penetration of flux in the sample. As a matter of fact, the modification of the above seen Meissner state profiles becomes less pronounced as the applied field H_a decreases and so is the penetration in the sample. For further physical insight the reader is invited to check the corresponding lines of magnetic flux density **B**, which are fully equivalent to those obtained for the disk geometry and shown in figure 9.9.

Inverse problem
The 'inversion' of experimental data, i.e. the 'manipulation' of $B_z(x')$ so as to obtain the underlying current density profile $K(x)$ has been a milestone in MOI studies [3]. In brief, the idea is to straightforwardly derive the current density profile from the experimental observations, by contrast to using model calculations based on different ansatzs for K and comparing to the real facts. In general, this is a difficult problem, many times mathematically ill-posed and it needs dedicated consideration for each case. Just consider that for an arbitrary situation one would need to know (B_x, B_y, B_z) over some region of space in order to have a unique solution for the

---

[3] In this case, the sheet current is the average across the strip: $K(x) = \int J_y(x, z)dz$.
[4] Notice the analogy to Mikheenko's model for thin discs in equation (9.11).

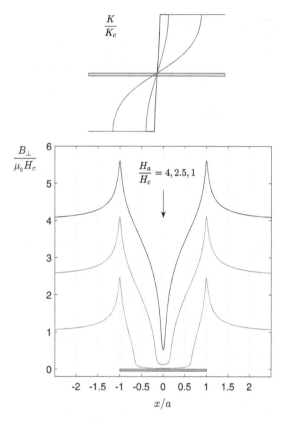

Figure 11.6. Total flux distribution along the dashed line over a long strip (figure 11.1) in the critical state for the height $h = 0.02a$ and different values of the applied magnetic field H_a (see text for units). Also shown are the sheet current profiles. 🔗. Source codes available at https://iopscience.iop.org/book/978-0-7503-2711-4.

unknown current distribution. Here, we will put forward an introductory scheme based on [7] that applies to quasi-1D geometries. As shown in that work, based on the linearity of the forward relation (equation (11.6)) one may use matrix inversion. Thus, if one discretizes the problem and writes

$$|\mathbf{B}_z^{\text{sc}}\rangle = \mathbf{Z}|\mathbf{K}\rangle, \tag{11.7}$$

one may find an inverse relation of the kind

$$|\mathbf{K}\rangle = \mathbf{R}|\mathbf{B}_z^{\text{sc}}\rangle \tag{11.8}$$

The authors propose a 'filtered' version that smoothes oscillations near the points where $K(x)$ changes abruptly, with the inversion matrix given by

$$R(i, j) = \frac{i - j}{\pi} \left\{ \frac{1 - (-1)^{i-j} e^{\pi d}}{d^2 + (i - j)^2} + \frac{[d^2 + (i - j)^2 - 1][1 + (-1)^{i-j} e^{\pi d}}{[d^2 + (i - j + 1)^2][d^2 + (i - j - 1)^2]} \right\}$$

Here, the indices i, j discretise the points for the spatial dependencies, i.e. $x = i\delta$, $x' = j\delta$, $h = d\delta$.

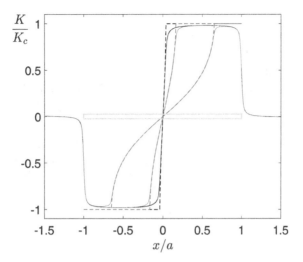

Figure 11.7. Sheet currents recovered by application of equation (11.8) to the $B_z(x')$ data of figure 11.6. The reference $K(x)$ profiles (dashed lines) are included for comparison. 🔗. Source codes available at https://iopscience.iop.org/book/978-0-7503-2711-4.

Figures 11.6, 11.7 and their extensions may be constructed with help of the following scripts:

- `brandt_indenbom_model.m` (obtain $K(x)$ from equation (11.4))
- `recover_K.m` (perform inversion through equation (11.8))

The results of applying the inversion scheme to the configurations of figure 11.6 are shown in figure 11.7. From the simulated $B_z(x')$ profiles[5], we straightforwardly obtain $K(x)$ and compare to the exact values calculated with equation (11.4). As the reader may check by using the scripts provided, the closer to the sample (smaller values of the distance h), the better the performance of the inversion formula.

11.2.2 Thin sample limit: square platelet

Eventually, we introduce some ideas on the performance of MOI experiments above superconducting platelets in the critical state. In this case we limit ourselves to the forward problem, i.e. assume a model for the critical state and predict the profiles of B_z. Specialised techniques for the inverse problem that generalise the quasi-1D limit in equation (11.8) to 2D problems may be found in the literature [3, 8].

The prediction of the flux profiles above a superconducting platelet under a uniform applied field may be done by using the methods introduced in section 10.2. Recall that, taking advantage of the *stream function* formalism, we introduced a method to obtain $B_z(x, y)$ over the square platelet. Thus, for a given critical state of the sample, characterised by the solution $\sigma(x, y)$, and by means of equation (10.16)

---

[5] It must be noted that inversion is to be applied to the magnetic field created by the superconductor. In the case of using experimental data, one should not forget to subtract the contribution related to the external applied field.

one may map B_z over the sample. Here, we generalise that result in order to simulate the scan of field profiles beyond the sample's surface. For this, one must simply reformulate equation (10.17) so as to account for the magnetic vector potential of the superconductor at points above the surface (recall that $|A_{x,y}\rangle = m_{x,y}\,D_{x,y}|\sigma\rangle$). Thus, we define augmented matrices, such that $|B_z^{sc}\rangle$ is obtained by using

$$\mathbf{Z} = -(\mathbf{D}_x^{aug}\,\mathbf{m}_x^{aug}\,\mathbf{D}_x + \mathbf{D}_y^{aug}\,\mathbf{m}_y^{aug}\,\mathbf{D}_y) \qquad (11.9)$$

$\mathbf{Z}$ will be a non-square matrix acting on the points of the superconductor (x, y) and giving values to the field at the desired points (x', y'). As suggested by the above equation, a technical detail of mention is that, in general, the calculation of B_z requires the computation of both A_x and A_y. Analytically

$$B_z = \frac{\partial A_y}{\partial x} - \frac{\partial A_x}{\partial y} \qquad (11.10)$$

This implies that, theoretically, obtaining B_z along a given line generally implies to evaluate A_x, A_y in additional points so as to allow the calculation of derivatives.

Figure 11.8 shows the results for B_z along two lines parallel to the x-axis. We call the readers' attention that the profile along the central line $(x', y' = 0)$ strongly

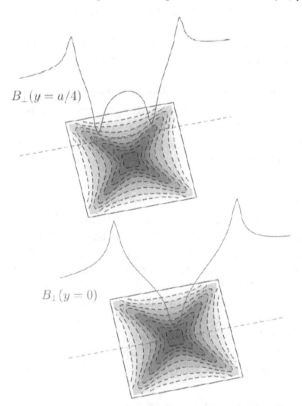

Figure 11.8. Flux distribution along the dashed lines over the square platelet $(a \times a)$ of figures 10.6 and 10.7 for the magnetisation state labelled B. ✐. Source codes available at https://iopscience.iop.org/book/978-0-7503-2711-4.

resembles that of long samples (just compare to the case $H_a = 4H_c$ in figure 11.6). However, one must notice that the scan along the parallel line $(x', y' = a/4)$ gives a rather different result. As noticeable in the underlying colour map, the fully penetrated state is characterised by the 'cushion-like' structure of B_z and this is clearly shown by the increase around the central part of the second scan.

In order to create the plots in figure 11.8, we have used the scripts

- `MOI_platelet.m`
- `Figure_10_7_like.m`

They are built by combination of the scripts provided in figure 10.7 with a number of modifications that are specified in the source codes.

11.3 Review exercises and challenges

11.3.1 Review exercises

Review exercise R11-1

Reproduce figure 11.5 for the case of an inclined garnet above the superconductor, i.e. study the effect of a misorientation of the garnet and the superconductor (see figure 11.9). For quantitative evaluation suppose that the superconductor is a beam of cross section $700\ \mu m \times 70\ \mu m$ and that the garnet is at a distance $15\ \mu m$ above the left edge, while at $30\ \mu m$ above the right one.

Hint: evaluate the component of the magnetic field perpendicular to the surface of the garnet in terms of the B_x, B_z components, that are obtained by application of Ampère's law (equation (11.6) for B_z and alike for B_x).

Review exercise R11-2

Reproduce figure 11.5 for the case of a superconducting disk (thin cylinder) in the Meissner state. For this, use the scripts related to the problem in figure 9.3. Evaluate the field along the radial direction above the sample and plot $B_z(0 < r < 1.5R, \varphi = (0, \pi), z_0)$

Hint: use the scripts available in order to obtain the vector potential $A_\varphi(r, z_0)$ and recall

$$B_z = \frac{1}{r}\frac{\partial}{\partial r}\left(rA_\varphi\right) \tag{11.11}$$

Apply finite differences as done in figure 11.8 for obtaining derivatives.

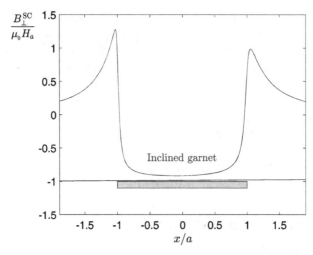

Figure 11.9. Distribution of $B_\perp^{sc}$ across the width of the superconducting beam of the aspect ratio $\zeta = 0.1$ with an inclined garnet on top (see exercise R11-1).

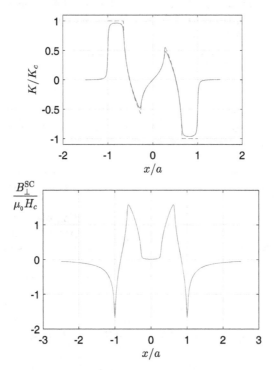

Figure 11.10. Lower: simulated profile of perpendicular flux above the superconducting strip for $H_0 = 2H_c$ and $H_a = 0$. Upper: the current density profile in the simulated forward problem (dashed) and recovered (continuous) by applying equation (11.8). (Exercise C11-1)

Review exercise R11-3

Reproduce figure 11.6 for the case of a superconducting disk (thin cylinder) in the critical state. Evaluate the field along the radial direction above the sample and plot $B_z(0 < r < 1.5R, \varphi = (0, \pi), z_0)$

Hint: use the scripts related to the problem in figure 9.7 in order to obtain the vector potential $A_\varphi(r, z_0)$ as in the previous exercise.

Review exercise R11-4

By using equations (11.4) and (11.6), show that one may obtain analytical expressions for the perpendicular magnetic field [6]

$$B_z(x) = \begin{cases} 0 & ; \quad |x| < b_B \\[2ex] \mu_0 H_c \tanh^{-1}\left[\dfrac{\sqrt{x^2 - b_B^2}}{c_B|y|}\right] & ; \quad b_B \leqslant |x| \leqslant a \\[2ex] \mu_0 H_c \tanh^{-1}\left[\dfrac{c_B|y|}{\sqrt{x^2 - b_B^2}}\right] & ; \quad |x| > a \end{cases} \qquad (11.12)$$

Check them against the profiles obtained by numerical integration of the current density.

11.3.2 Challenges

Exercise C11-1

Complete figures 11.6 and 11.7 with the simulation of field profiles and recovered current density in the *reversal* critical state of the long strip. Assume that a maximum applied field H_0 has been reached and then evaluate $B_z(x')$ and $K(x)$ corresponding to a return field $H_a < H_0$. See figure 11.10 as an orientation.

Hint: according to [6] the reversal process may be evaluated from

$$K_\searrow(x, H_a, K_c) = K(x, H_0, K_c) - K(x, H_0 - H_a, 2K_c) \qquad (11.13)$$

References

[1] Bending S J 1999 Local magnetic probes of superconductors *Adv. Phys.* **48** 449–535
[2] Johansen T H and Shantsev D V 2004 *Magneto-Optical Imaging. II. Mathematics, Physics and Chemistry* Nato Science Series 01 vol 142 (Dordrecht: Springer)

[3] Jooss C, Albrecht J, Kuhn H, Leonhardt S and Kronmüller H 2002 Magneto-optical studies of current distributions in high-T_c superconductors *Rep. Prog. Phys.* **65** 651–788

[4] Brandt E H and Mikitik G P 2000 Meissner-London currents in superconductors with rectangular cross section *Phys. Rev. Lett.* **85** 4164–7

[5] Grisolia M N, Badía-Majós A and van der Beek C J 2013 Imaging flux distributions around superconductors: geometrical susceptibility in the meissner state *J. Appl. Phys.* **114** 203904

[6] Brandt E H and Indenbom M 1993 Type-II-superconductor strip with current in a perpendicular magnetic field *Phys. Rev.* B **48** 12893–906

[7] Johansen T H, Baziljevich M, Bratsberg H, Galperin Y, Lindelof P E, Shen Y and Vase P 1996 Direct observation of the current distribution in thin superconducting strips using magneto-optic imaging *Phys. Rev.* B **54** 16264–9

[8] Jooss C, Warthmann R, Forkl A and Kronmüller H 1998 High-resolution magneto-optical imaging of critical currents in $YBa_2Cu_3O_{7-\delta}$ thin films *Physica C: Supercond. Appl.* **299** 215–30

IOP Publishing

Macroscopic Superconducting Phenomena
An interactive guide
Antonio Badía-Majós

Chapter 12

Interaction with magnets: force microscopies

Continuing along the line established in the previous chapter, here we will provide analytical and numerical tools related to the extraction of physical parameters from the interaction between a magnetic material and a superconductor. Specifically, we deal with a technique developed in the early 1990s as reported by Bending [1], i.e. magnetic force microscopy[1]. As said in section 4.4 this technique has been successfully used to obtain superconducting parameters as the London's penetration depth λ_L and also applied to manipulate individual vortices in superconductors. Typical experiments with MFM tips scan over regions of submillimeter size, thus allowing to extract local information on the sample with resolution in the nm scale. Nevertheless, the interaction of the tip and the underlying superconductor is customarily described in the language of the macroscopic Maxwell equations (though microscopic variations are involved), which will be the focus of this chapter. Essentially, we will introduce methodological concepts to obtain the electromechanics of magnet-superconductor systems. Here, the tip will be considered a uniformly magnetised object, and the superconductor characterised by the Meissner state response. Partial penetration of the magnetic field parameterised by λ_L will be allowed.

The chapter is organised in two parts. To start with, we present the so-called *forward problem*, i.e. the prediction of the repulsion force in terms of geometry and λ_L. In the second part, we afford the inverse problem. This entails to introduce the systematics that allows to extract the superconducting property, i.e. λ_L in terms of the measured repulsion force.

12.1 Forward problem: prediction of the force

Let us begin with the justification of the physical model that reasonably describes the experiments. A typical range for the distance between the magnetic tip and the

---

[1] In what follows, we use MFM for magnetic force microscopy.

doi:10.1088/978-0-7503-2711-4ch12

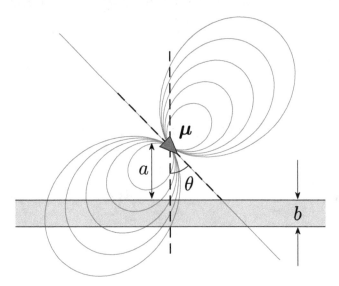

Figure 12.1. Sketch of the MFM configuration. The magnetic tip settles at a distance a and may be tilted through an angle θ above the superconducting film of thickness b that occupies the region $-b \leqslant z \leqslant 0$.

superconductor is $a \lesssim 1\mu$m and the size of the tip around $0.5\,\mu$m. On the other hand, sample dimensions may be in the mm range, while thickness depends on the sample, but could be of the order of λ_L. Thus, a reasonable model considers a finite size magnet above a superconducting slab (see figure 12.1) or sometimes above a semi-infinite superconductor (in case that $b \gg \lambda_L$). In fact, as recalled in [2], neglecting the size of the magnet, i.e. treating it as a point dipole, may lead to overestimating the penetration depth of the superconductor in several times its value.

In this section we show how to predict the force measured in terms of λ_L by discussing the solution of the field equations for **B**. This means to solve the Maxwell and London equations for the given geometry. A collection of numerical resources for further investigation of related problems is provided.

12.1.1 Point dipole approximation

To start with, we put forward some expressions related to the magnetic field induction that solves the problem of a point dipole above a superconducting film in the Meissner state.

The problem may be stated as follows [3]. We split **B(r)** in the form

$$
\mathbf{B}(\mathbf{r}) = \begin{cases} \mathbf{B}_1(\mathbf{r}) + \mathbf{B}_2(\mathbf{r}) & ; \quad z \geqslant 0 \\ \mathbf{B}_3(\mathbf{r}) & ; \quad -b \leqslant z \leqslant 0 \\ \mathbf{B}_4(\mathbf{r}) & ; \quad z \leqslant -b \end{cases} \tag{12.1}
$$

Above, $\mathbf{B}_1$ is the field created by the source dipole, $\mathbf{B}_2$ the contribution of the Meissner currents above the superconductor, $\mathbf{B}_3$ the total field penetrating the sample, and $\mathbf{B}_4$ the field beneath. By analogy to the problem of wave scattering across a barrier, these fields will be named after *incident*, *reflected*, *penetrating* and *transmitted*, respectively.

Without going into detail, we put forward the solution for these fields[2]. Considering that rotational symmetry will be often recalled, the fields are conveniently expanded in terms of Bessel functions (associated *wavenumber k*), with the coefficients obtained by imposing boundary conditions at the interfaces [3].

As customary, the dipole field is written in the form

$$\mathbf{B}_1(\mathbf{r}) = \frac{\mu_0}{4\pi}\left[\frac{3\mathbf{r}'(\boldsymbol{\mu}\cdot\mathbf{r}')}{r'^{\,5}} - \frac{\boldsymbol{\mu}}{r'^{\,3}}\right] \tag{12.2}$$

with $\mathbf{r} \equiv (x, y, z)$, $\mathbf{r}' \equiv (x, y, z - a)$ if the dipole settles at the point $(0, 0, a)$.

On the other hand, one finds

$$B_{2_i} = \frac{\mu_0}{4\pi}\int_0^\infty dk\left[C_{2_{xy}}(k)k^2 e^{-k(a+z)}\left(\sum_j G_{ij}m_j\right)\right]_{i=x,y}$$

$$B_{2_i} = \frac{\mu_0}{4\pi}\int_0^\infty dk\left[C_{2_z}(k)k^2 e^{-k(a+z)}\left(\sum_j G_{ij}m_j\right)\right]_{i=z}$$

$$B_{3_i} = \frac{\mu_0}{4\pi}\int_0^\infty dk\left[\left(C_{3_{xy}}^+(k)e^{\gamma z} + C_{3_{xy}}^-(k)e^{-\gamma z}\right)k^2 e^{-ka}\left(\sum_j G_{ij}m_j\right)\right]_{i=x,y}$$

$$B_{3_i} = \frac{\mu_0}{4\pi}\int_0^\infty dk\left[\left(C_{3_z}^+(k)e^{\gamma z} + C_{3_z}^-(k)e^{-\gamma z}\right)k^2 e^{-ka}\left(\sum_j G_{ij}m_j\right)\right]_{i=z} \tag{12.3}$$

$$B_{4_i} = \frac{\mu_0}{4\pi}\int_0^\infty dk\left[C_{4_{xy}}(k)k^2 e^{-k(a-z-b)}\left(\sum_j G_{ij}m_j\right)\right]_{i=x,y}$$

$$B_{4_i} = \frac{\mu_0}{4\pi}\int_0^\infty dk\left[C_{4_z}(k)k^2 e^{-k(a-z-b)}\left(\sum_j G_{ij}m_j\right)\right]_{i=z}$$

where we have used: (i) $\gamma = \sqrt{k^2 + 1/\lambda_L^2}$, (ii) the geometrical matrix elements G_{ij} that appear by writing the field in cylindrical functions, and the coefficients $C(k)$ determined by the boundary conditions. Their respective expressions are

---

[2] Just recall that $\mathbf{B}_2$, $\mathbf{B}_4$ satisfy Laplace's equation, while $\mathbf{B}_3$ satisfies London's equation.

$$G_{11} = -\frac{1}{2}J_0(k\sqrt{x^2 + y^2}) + \frac{x^2 - y^2}{2(x^2 + y^2)}J_2(k\sqrt{x^2 + y^2})$$

$$G_{12} = \frac{xy}{x^2 + y^2}J_2(k\sqrt{x^2 + y^2})$$

$$G_{13} = -\frac{x}{\sqrt{x^2 + y^2}}J_1(k\sqrt{x^2 + y^2})$$

(12.4)

$$G_{22} = -\frac{1}{2}J_0(k\sqrt{x^2 + y^2}) + \frac{y^2 - x^2}{2(x^2 + y^2)}J_2(k\sqrt{x^2 + y^2})$$

$$G_{23} = -\frac{y}{\sqrt{x^2 + y^2}}J_1(k\sqrt{x^2 + y^2})$$

$$G_{33} = J_0(k\sqrt{x^2 + y^2})$$

and

$$C_{2_{xy}} = -C_{2_z} = \Delta(\gamma^2 - k^2)(1 - e^{-2\gamma b})$$

$$C_{3_{xy}}^+ = 2\gamma\Delta(\gamma + k)$$

$$C_{3_{xy}}^- = -2\gamma\Delta(\gamma - k)e^{-2\gamma b}$$

$$C_{3_z}^+ = 2k\Delta(\gamma + k)$$

$$C_{3_z}^- = 2k\Delta(\gamma - k)e^{-2\gamma b}$$

$$C_{4_{xy}} = C_{4_z} = 4k\gamma\Delta e^{-\gamma b}$$

(12.5)

with the definition $\Delta \equiv 1/[(\gamma + k)^2 - (\gamma - k)^2 e^{-2\gamma b}]$.

The above expressions may be used for several purposes. First, we apply them in order to obtain the picture of magnetic field lines related to the MFM problem under various conditions. Figure 12.2 displays the lines of **B** for different orientations of the magnetic tip and different values of the penetration depth. In this case, the 2D picture uses $\boldsymbol{\mu} = m_0(0, \cos\theta, -\sin\theta)$, which introduces some simplification in equation (12.3). The field lines have been straightforwardly generated with the help of the MATLAB built-in function streamlines() that inputs the components of the vector field **B(r)** at some collection of grid points and integrates the differential relation between **B** and the tangent vector along the lines $d\ell$. In the 2D statement this means

$$\mathbf{B}\|d\ell \Rightarrow \frac{dz}{dy} = \frac{B_z}{B_y} \Rightarrow \int dz = \int \frac{B_z}{B_y}dy$$

(12.6)

Figure 12.2 has been constructed with help of the following scripts:
- figure_12_2.m (main module)
- incident_exact.m (evaluate incident field $\mathbf{B}_1$)
- reflected.m (evaluate reflected field $\mathbf{B}_2$)
- penetrating.m (evaluate penetrating field $\mathbf{B}_3$)
- transmitted.m (evaluate transmitted field $\mathbf{B}_4$)
- m_point.m (plot oriented magnetic tip)

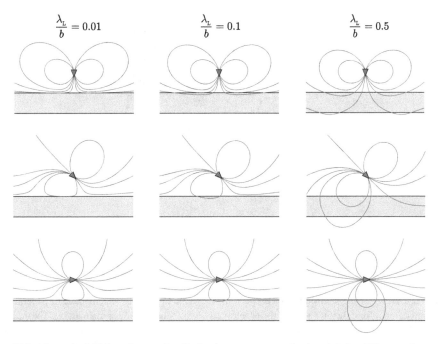

Figure 12.2. Magnetic field lines for a point dipole above a superconducting slab for different values of the penetration depth and for different tilt angles of the dipole ($\theta = 0$, $\pi/4$, $\pi/2$). 🔗. Source codes available at https://iopscience.iop.org/book/978-0-7503-2711-4.

Another quantity of interest is the actual force between the tip and the superconductor. Within the infinite slab approximation, the horizontal translational symmetry imposes $F_x = F_y = 0$. The vertical force, on the other side, may be evaluated from the self-interaction energy $U = -(1/2)\boldsymbol{\mu} \cdot \mathbf{B}_2$.[3] Then, starting with equation (12.3), and evaluating at $x, y \to 0$ one gets

$$F_z = -\frac{\partial U}{\partial a} = \frac{\mu_0}{8\pi} \int_0^\infty dk \; C_{2_{xy}}(k) k^3 e^{-2ak}(m_0^2 + m_z^2) \tag{12.7}$$

This expression has been used to produce figure 12.3, that relates to the configuration $\theta = 0$ in figure 12.2. This is implemented in the script `force_dipole_mfm.m`. Normalised units in the plot are defined by

$$f\left(\frac{a}{b}, \frac{\lambda_{\mathrm{L}}}{b}\right) \equiv \frac{1}{1 + \cos^2\theta} \frac{8\pi b^4}{\mu_0 m_0^2} F_z(a, b, \lambda_{\mathrm{L}}) \tag{12.8}$$

---

[3] We recall that the factor 1/2 is present because $\mathbf{B}_2$ is not given but induced by the presence of the magnet itself (see section 13.1.1 for further details).

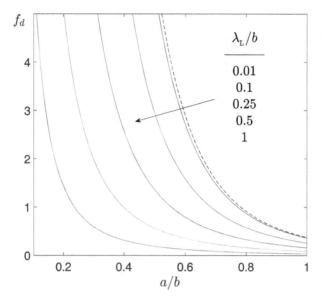

Figure 12.3. Decay of the repulsion force between the magnetic tip and the superconductor in terms of the separation distance for different values of the penetration depth (point dipole approximation). The dashed line corresponds to the limiting case $\lambda_L = 0$. Vertical scale is multiplied by 10^6 and normalised units are used (see text). 🔗. Source codes available at https://iopscience.iop.org/book/978-0-7503-2711-4.

12.1.2 Finite size effects

For many purposes, equation (12.7) must be generalised in view of the realistic consideration that the tip is spatially extended. This can be done by calling on superposition. Thus, if one uses $\mathbf{M}(\mathbf{r})$ for the magnetisation function of the tip and V for its volume, one gets the general expression

$$
F_z = \frac{\mu_0}{4\pi} \int_0^\infty dk \int_V d^3\mathbf{r} \int_V d^3\mathbf{r}'
$$
$$
\times \left[C_{2xy}(k) k^3 e^{-k(2a+z+z')} \left(\sum_{i,j} M_i(\mathbf{r}) g_{ij}(\mathbf{r} - \mathbf{r}') M_i(\mathbf{r}') \right) \right]
\tag{12.9}
$$

with $g_{ij} = \pm G_{ij}$ for either $i \neq j$ or $i = j$.

The above formula must be specified for the tip's geometry. Below, we introduce the realistic situation of using a conical tip as well as some indications that allow to afford the involved multiple integration (septuple in general) with relative ease.

First, we notice that upon the assumption of a uniformly magnetised tip ($\mathbf{M} = M_0 \hat{k} \equiv m_0 \hat{k}/V$), the bracketed expression in (12.9) simplifies to $-M_0^2 G_{33}(\mathbf{r} - \mathbf{r}')$. Then equation (12.9) becomes

$$
F_z = \frac{\mu_0 M_0^2}{4\pi} \int_0^\infty dk \int_V dx\, dy\, dz \int_V dx'dy'dz'
$$
$$
\times \left[\frac{(1 - e^{-2\gamma b})/\lambda_L^2}{(\gamma + k)^2 - (\gamma - k)^2 e^{-2\gamma b}} k^3 e^{-k(2a+z+z')} J_0(k\sqrt{(x - x')^2 + (y - y')^2}) \right]
\tag{12.10}
$$

Still, this is a cumbersome expression that may further disentangled by using cylindrical coordinates and recalling the properties of Bessel functions [4, 5]. One gets

$$F_z = \mu_0 M_0^2 \pi \int_0^\infty dk \int_0^R d\rho \int_0^R d\rho' \frac{(1 - e^{-2\gamma b})/\lambda_L^2}{(\gamma + k)^2 - (\gamma - k)^2 e^{-2\gamma b}} k \qquad (12.11)$$
$$\times [\rho\rho' J_0(k\rho) J_0(k\rho')(e^{-k\rho h/R} - e^{-kh})(e^{-k\rho' h/R} - e^{-kh})]$$

a triple integral that may be straightforwardly solved by means of the MATLAB function integral3().

Another possibility to tackle with the statement in equation (12.10) is to invoke series expansion. Thus, considering that the function $C_{2_{xy}}(k)$ 'codifies' the physical information related to the penetration depth and that the *wavenumber k* is gauged in terms of a by the exponential factor, one may power expand the integrand in terms of $\lambda_L k$ and rescale k by $1/a$. Formally, this leads to the following expression for F_z

$$F_z(\lambda_L, a) = F_0(a) + \sum_{n=1}^\infty F_n(a)\left(\frac{\lambda_L}{a}\right)^n \qquad (12.12)$$

where $F_0(a)$ represents the value of the force in the 'perfect' Meissner limit (bulk approximation), and $F_n(a)$ are geometry dependent correction functions. Remarkably, these functions may be obtained by using recurrence relations. In the thick sample limit ($b \gg a, \lambda_L$) one has [6]:

$$F_n(a) = \alpha_n\left(\frac{-1}{2}\right)^n\left(\frac{d^n F_0}{da^n}\right) a^n \qquad (12.13)$$

with

$$\alpha_1 = -\alpha_2 = -2, \ \alpha_3 = -1, \ \alpha_{2m+1} = \frac{(-1)^m (2m - 3)!!}{2^{m-1} m!} \qquad (12.14)$$

Thus, theoretically, one may evaluate the force upon knowledge of the limiting behaviour $F_0(a)$ for the ideal Meissner state ($\lambda_L = 0$). Notice that $F_0(a)$ may be obtained from (12.11) just by replacing $C_{2_{xy}} \rightarrow 1$.

Figure 12.4 shows the function $F_z(a)$ for a conical magnetic tip magnetised along the axis and with aspect ratio $R/h = 0.25$ above a superconducting film for several values of the penetration length. By comparing to figure 12.3, one may notice the importance of considering the finite size of the tip, as well as the thickness of the superconductor (if comparable to λ_L). Here, normalised units are defined by

$$f_z \equiv \frac{2\pi^2 R^2 h^2}{9\mu_0 m_0^2} F_z \qquad (12.15)$$

Figure 12.4 is built by applying equation (12.11), that gives the MFM force on a conical tip. It relies on the scripts
- figure_12_4.m (main module)
- force_mfm_cone.m
- plot_cone.m (plots the tip).

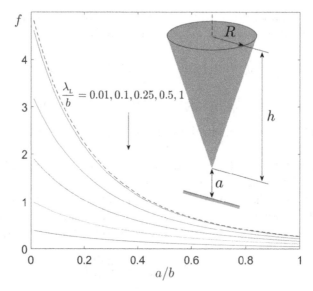

Figure 12.4. Decay of the repulsion force between a conical magnetic tip and a superconducting film of thickness b for different values of the penetration depth as calculated with equation (12.11). The dashed line corresponds to the limiting case $\lambda_L = 0$. Vertical scale is multiplied by 10^5 and normalised units are used (see text). $\mathscr{P}$. Source codes available at https://iopscience.iop.org/book/978-0-7503-2711-4.

12.2 Inverse problem: prediction of λ_L

The determination of the penetration depth from MFM experiments may be done by using λ_L as a fit parameter within the above presented *forward* scheme. Thus, by optimising the agreement between the calculated theoretical force and the experimental data one may obtain the expected value. This procedure was used in [2] with excellent results for a Nb film. In fact, the authors considered the field $\mathbf{B}_2$ above individual vortices within the sample for the calculation of the force by a procedure that is fully parallel to the method in section 12.1.2[4]. However, one must mention that this technique requires an excellent knowledge of the experiment's geometry in order to have a single fit parameter, i.e. λ_L. Alternatively, one may use a so-called *inversion procedure* that straightforwardly manipulates the experimental data. The focus of such methods is to avoid fitting operations with multiple parameters that may have an ambiguous interpretation. Below, we elaborate this concept for the MFM problem. We will concentrate on how to cope with the uncertainty related to the geometry of the tip. Other issues that may be overcome by the introduction of inverse methods have been reported in the literature. For instance, one may treat problems with inhomogeneity in the material (spatially dependent $\lambda_L^2(\mathbf{r})$ as shown in [8]).

---

[4] The presence of vortices may introduced by using a modified London's equation within the film (field $\mathbf{B}_3$ in equation (12.1)), that models the vortex through an impulse function [7] (see also exercise C12-1).

We start by stepping back to equation (12.12). By rewriting it in the form

$$F_z(\lambda_L, a) - F_0(a) = \sum_{n=1}^{\infty} F_n(a)\left(\frac{\lambda_L}{a}\right)^n \tag{12.16}$$

and defining $\bar{F}(\lambda_L, a) \equiv F_z(\lambda_L, a) - F_0(a)$ one may state the inverse problem as find the coefficients of the inverse expansion

$$\lambda_R(a) = \sum_{n=1}^{\infty} c_n \bar{F}(\lambda_L, a)^n \tag{12.17}$$

Here, one must notice that $\bar{F}(\lambda_L, a)$ means the *experimental* quantity that one may obtain from the actual measurements on the sample $F_z(\lambda_L, a)$, and the *calibration* function $F_0(a)$, which must be obtained for the actual tip[5]. Relying on the use of complex variables (Cauchy's integral formula [5]) we obtain the general expression

$$c_n = \lim_{\lambda_L \to 0} \frac{1}{n!}\left[\frac{d^{n-1}}{d\lambda_L^{n-1}}\frac{\lambda_L^n}{\bar{F}^n}\right] \tag{12.18}$$

Then, taking advantage of relations (12.12)–(12.14) one may find that these coefficients may be written in terms of the calibration function $F_0(a)$ and its derivatives. For the readers' sake, several of them are collected in table 12.1.

In order to check the stability and effectiveness of the method one may perform a simulation for synthetic data $\bar{F}(\lambda_L, a)$. Figure 12.5 shows the reconstruction of the penetration depth based on the theoretical data for the conical tip in figure 12.4. Here, we choose the value $(\lambda_L = 0.1b)$ and apply the expansion (12.17) for increasing number of terms. One may notice that four terms give a rather reasonable reconstruction, and also the asymptotic character of the dependence of the recovered penetration depth $\lambda_R(a)$. The recovery is performed by applying equation (12.17) to a set of synthetic data produced by equation (12.11). It uses the script `force_mfm_recover.m`.

Table 12.1. Coefficients for the inversion series in terms of the measured MFM force, as in equation (12.17).

Coefficient	Value
c_1	$1/F_0'$
c_2	$-(1/2)/F_0''/(F_0')^3$
c_3	$(1/8)[4(F_0'')^2 - F_0'F_0^{(3)}]/(F_0')^5$
c_4	$(5/8)[(1/2)F_0'F_0''F_0^{(3)} - (F_0'')^3]/(F_0')^7$

---

[5] Experimentally, one may obtain $F_0(a)$ either by measuring on a reference superconductor under conditions that allow to neglect the penetration depth $(\lambda_L \to 0)$ or also a high permeability material $(\mu \to \infty)$.

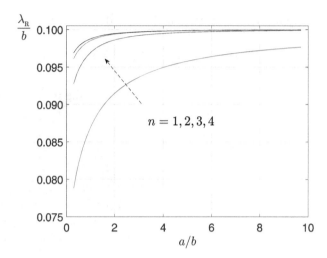

Figure 12.5. Reconstruction penetration depth λ_R (theoretical $\lambda_L/b = 0.1$) by means of equation (12.17) for a simulated force versus distance experiment in the case of a conical tip (see figure 12.4). As indicated, the reconstruction is done by an increasing number of coefficients $n = 1, 2, 3, 4$. 🔗. Source codes available at https://iopscience.iop.org/book/978-0-7503-2711-4.

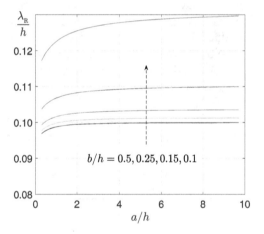

Figure 12.6. Influence of the thickness of the superconductor in the reconstructed penetration depth λ_R (theoretical $\lambda_L/h = 0.1$) for the case of a conical tip (see figure 12.4) and by using number of 4 coefficients in equation (12.17). We use the height of the tip as the normalisation distance. 🔗. Source codes available at https://iopscience.iop.org/book/978-0-7503-2711-4.

Another aspect that must be considered is that equation (12.17) was derived for the limiting case of very thick samples ($b \gg \lambda_L$). In order to establish the range of validity, we put forward figure 12.6 and the related software (`force_mfm_recover_b.m`). Here, we have simulated the forward problem for several values of increasing λ_L and apply the 'thick sample' recovery equation (12.17). As one may recall in the figure, the inversion method works well for $b \gtrsim 5\lambda_L$.

12.3 Review exercises and challenges

12.3.1 Review exercises

Review exercise R11-1

The coefficient $C_{2_{xy}}(k)$ in equation (12.5) is of importance because for standard configurations (tip uniformly magnetised along the normal to the plane of the sample) it determines the MFM force (see equation (12.9) and subsequent). Show that for thick samples, one may approximate it by

$$C_{2_{xy}}(k) \xrightarrow[b \gg a, \lambda_L]{} \frac{\sqrt{1 + \lambda_L^2 k^2} - \lambda_L k}{\sqrt{1 + \lambda_L^2 k^2} + \lambda_L k} \qquad (12.19)$$

Hint: consider the role of b, a and λ_L. Numerical comparison of the force obtained from both expressions may be reassuring.

Review exercise R11-2

Find an expression equivalent to (12.11), but for the case of cylindrical and hemispherical tips with their axis perpendicular to the plane of the superconductor. Both shapes can be combined to simulate the case of a cylindrical tip with rounded end.

Hint: pose the problem in cylindrical coordinates and use the expression

$$J_0\left(k\sqrt{\rho^2 + \rho'^2 - 2\rho\rho' \cos\phi}\right) = \sum_{m=-\infty}^{\infty} e^{im\phi} J_m(k\rho) J_m(k\rho') \qquad (12.20)$$

Review exercise R11-3

Investigate the validity of the forward series approach (equation (12.12)) by comparison to the exact expression (equation (12.11)). In particular, consider the range of values λ_L/b and the number of terms used in the series. Recall that the series expansion is valid for thick samples

---

Hint: use the software related to figure 12.4. For the numerical derivatives, use the finite difference approximation as in the script of figure 12.5.

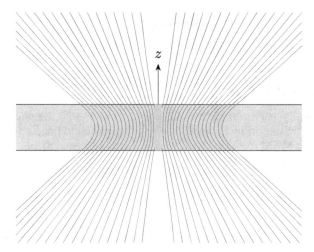

Figure 12.7. Magnetic induction lines created by a single vortex in a superconducting layer. The relation between the penetration depth and the thickness of the layer is $\lambda_L/b = 0.1$.

12.3.2 Challenges

Exercise C12-1 (MFM above individual vortices)

High resolution MFM above individual vortices has allowed to extract λ_L with precision [2] based on both vertical and lateral scans above the vortex. Related to this, and for the geometry sketched in figure 12.7:

(i) Show that the stray field of the vortex in a film of thickness b is given by

$$b_{1z}(\rho, z) = \int_0^\infty dk\ D_{1_z}(k)ke^{-kz}J_0(k\rho)\quad ;\quad z > 0$$

$$b_{2z}(\rho, z) = \int_0^\infty dk\ [D_{2_z}^+(k)\cosh(\gamma z) + D_{2_z}^-(k)\sinh(\gamma z)]kJ_0(k\rho)$$

$$+ \frac{\Phi_0}{2\pi\lambda_L^2}K_0(\rho/\lambda_L)\qquad\quad ; -b < z < 0 \tag{12.21}$$

$$b_{3z}(\rho, z) = \int_0^\infty dk\ D_{3_z}(k)ke^{k(z+b)}J_0(k\rho);\quad z < -b$$

with the definitions:

$$D_{1_z}(k) = D_{3_z}(k) \equiv \gamma\Delta_v\{\gamma\sinh(b\gamma) + k[\cosh(b\gamma) - 1]\}$$
$$D_{2_z}^+(k) \equiv -k\Delta_v\{k\sinh(b\gamma) + \gamma[\cosh(b\gamma) + 1]\}$$
$$D_{2_z}^-(k) \equiv -k\Delta_v\{\gamma\sinh(b\gamma) + k[\cosh(b\gamma) - 1]\}$$

and

$$\Delta_v \equiv \frac{\Phi_0/2\pi\gamma^2\lambda_L^2}{2k\gamma\cosh(b\gamma) + (k^2 + \gamma^2)\sinh(b\gamma)} \tag{12.22}$$

(ii) Obtain the radial components of the field by recalling its solenoidal character ($\partial_\rho(\rho b_\rho) = -\rho \partial_z b_z$ in this case) and recalling the recurrence relations for Bessel functions. Verify this by making a plot like figure 12.7.

(iii) Assuming that the vortex is pinned and that the planar dimensions of the superconductor are big enough, show that the lateral force on a small enough tip that scans horizontally above the magnet is

$$F_\rho(a, \rho) = -m_z \int_0^\infty dk \; D_{1_z}(k) k^2 e^{-ka} J_1(k\rho) \qquad (12.23)$$

with (a, ρ) being the position of the tip above the vortex, that settles at the origin of coordinates in the plane ($\rho = 0$). This expression allows to evaluate the behaviour of the force in a lateral scan above the vortex.

Hint: notice that the magnetic field satisfies Laplace's equation for $z > 0$ and $z < -b$, while one must impose the modified second London equation within the superconductor. This modification of equation (2.5) so as to include the presence of the vortex reads

$$-\lambda_\mathrm{L}^2 \, \nabla^2 \, \mathbf{B} + \mathbf{B} = \Phi_0 \delta(\mathbf{r} - \mathbf{r}_v)\hat{\boldsymbol{k}} \qquad (12.24)$$

when the vortex is along z-axis and settles at the point $\mathbf{r}_v$. Recall, that as suggested by the right pane of figure 2.3, this equation represents the limit of the Ginzburg–Landau equations for which one may use a constant order parameter for all space except for a singular points at $\mathbf{r} = \mathbf{r}_v$, and then solves the situation by using the London equation 'almost everywhere'.

As shown in [7], Fourier transforming in the plane XY-plane may be of help because of the translation symmetry of the sample, i.e. the actual position of the vortex is not relevant.

References

[1] Bending S J 1999 Local magnetic probes of superconductors *Adv. Phys.* **48** 449–535
[2] Nazaretski E, Thibodaux J P, Vekhter I, Civale L, Thompson J D and Movshovich R 2009 Direct measurements of the penetration depth in a superconducting film using magnetic force microscopy *Appl. Phys. Lett.* **95** 262502
[3] Badía A 2001 Inverse magnetic force microscopy of superconducting thin films *Phys. Rev.* B **63** 094502
[4] Jackson J D 1975 *Classical Electrodynamics* 2nd edn (New York: Wiley)
[5] Arfken G B and Weber H J 1995 *Mathematical Methods for Physicists* 4th edn (San Diego, CA: Academic)
[6] Badía A 1999 Asymptotic theory for the inverse problem in magnetic force microscopy of superconductors *Phys. Rev.* B **60** 10436–41
[7] de la Cruz de Oña A and Badía-Majós A 2004 Theory of vortex force microscopy in superconducting layers *Phys. Rev.* B **70** 144512
[8] Coffey M W 1999 Theory of inverse magnetic force microscopy of superconductors in half-space geometry *Phys. Rev. Lett.* **83** 1648–51

IOP Publishing

Macroscopic Superconducting Phenomena
An interactive guide
Antonio Badía-Majós

Chapter 13

Interaction with magnets: levitation

In the previous chapter we dealt with magnetic force microscopy problems. It was shown that, by analysing the force between a force microscopy tip and a superconductor, one may obtain information about the penetration depth of the material λ_L. Such experiments make sense in the microscopic scale, i.e. for samples of dimensions close to λ_L. In this chapter, we place the focus on quite the opposite scale, i.e. macroscopic systems for which $\lambda_L/L \to 0$ with L being the typical dimension of the superconductor. As the reader may guess, this relates to the macroscopic magnetic levitation applications that are already a fact. Generally speaking, we will deal with values of L in the scale of centimetres. As it will be shown below, by combination of such samples, one may reach levels of levitation force in the range of 10^3 N with relative ease.

For the readers' sake it must be mentioned that macroscopic physical models with a rather accurate predictive power are available to study these properties. An extensive and comprehensive compilation may be found in textbooks (see F Moon) [1] and review papers [2]. Roughly speaking, the different contributions may be classified as (i) *Meissner-limit* models and (ii) *critical-state* models. This chapter will be developed under such classification scheme. It divides into two sections. To start with, we will concentrate on the Meissner state limit, assuming complete flux expulsion within the superconductor. This ansatz will permit us to obtain various useful analytic approximations for the levitation force, though the problem will become numerical for the general case. In the second part of the chapter, we consider the levitation systems that include type-II superconductors with pinning. In general, this is a rather more interesting approach from the technological side, but becomes a rather less tractable problem from the mathematical point of view. It will require numerical methods for most situations.

The reader is provided with a collection of tools that allow to quantify several aspects of the technical problem: actual levitation forces, hysteresis phenomena, stability issues, etc. Although 3D modelling techniques are involved in the general

doi:10.1088/978-0-7503-2711-4ch13

statements, for which several calculation methods are available [3, 4], their level of sophistication is beyond our actual scope. Here, we will analyse rotationally and translationally symmetric systems, requiring at most 2D formulations. These are useful in several applications, such as rotating bearings, flywheels and motors, and transportation systems.

13.1 Levitation in the Meissner state

As will become clear, when the magnetic flux of a given source (typically permanent magnets) is expelled by a superconductor, one observes arising mechanical forces. A simple physical argument may be given. What occurs is that the induced super-currents' energy changes with the position of the sample in the field, and force appears so as to push it to the position with the lowest energy. Apparently, one may reciprocally think of the force on the magnet that will be opposite to the former. In fact, this principle will motivate a method of calculation in our exposition[1].

The above phenomenon was first reported in 1947 by V Arkadiev [5], who levitated a small magnet (4 × 4 × 10 mm) above a concave lead bowl in liquid helium. As will be suggested in exercise C13-1, the actual configuration used by Arkadiev obeys the physical requirement of achieving not only an equilibrium point in which gravity is compensated by the magnetic repulsion, but also a stable one. As must be shown in that exercise, the concave shape of the superconductor is essential for this.

Noteworthily, as will become apparent, macroscopic levitation systems seldom operate with superconductors in the Meissner state due to stability issues. However, for a given superconducting sample, the maximum attainable force corresponds to this regime. Below, we present several methods to evaluate such limit. They will be tested against more realistic models.

As said above, the non-uniformity of the magnetic field is an essential property with regards to the appearance of magnetic forces. For this reason, geometry and size effects will be determinant. Figure 13.1 displays the parameters for modelling the basic levitation experiment that concerns a cylindrical magnet above a cylindrical superconductor.

13.1.1 Meissner state: analytical solutions

In this section, we put forward a series of approximations valid for magnets of various dimensions and thin superconductors (**disk geometry**: $h_s/R_s \to 0$). In general, a finite thickness of the superconductor requires the use of numerical methods. Nevertheless, as it will be eventually shown, for many situations, the screening currents basically flow in the uppermost region of the superconducting cylinder, and the disk geometry gives a very good approximation to the problem.

---

[1] It is apparent that non-uniformity in the applied field is essential in order to observe magnetic force.

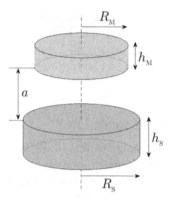

Figure 13.1. Sketch of the rotationally symmetric geometry considered in this chapter. A cylindrical magnet floats above a coaxial superconducting cylinder at the height a. Uniform magnetisation along the axis $\mathbf{M} = M_0\hat{k}$ will be assumed along this chapter. 🔗. Source codes available at https://iopscience.iop.org/book/978-0-7503-2711-4.

Small-nearby magnet (dipole–dipole approximation)
As sketched in figure 13.2, we start by considering the magnetic force acting on a tiny permanent magnet, close to the surface of an ideal superconducting sheet, i.e. R_s may be considered infinite.

If the dimensions of the magnet may be fully neglected, it can be described as a point dipole and one may take advantage of standard magnetostatics in order to evaluate the arising forces. Specifically, by using the image technique in electromagnetism [6], one may show that flux expulsion is achieved by introducing an *image dipole* as shown in the plot. Such image exactly replaces the action of the induced supercurrents in the upper half-space. Recall that this cancels the normal component of $\mathbf{B}$ on the sheet, a fact that ensures flux expulsion. Calculations proceed as follows.

We start by recalling the expression of the magnetic field created by a magnetic dipole

$$\mathbf{B}(\mathbf{r}) = \frac{\mu_0}{4\pi}\left[\frac{3\mathbf{r}(\boldsymbol{\mu}\cdot\mathbf{r})}{r^5} - \frac{\boldsymbol{\mu}}{r^3}\right] \tag{13.1}$$

Then one may apply the expression $\mathbf{F} = -\nabla(-\boldsymbol{\mu}\cdot\mathbf{B})$ so as to obtain the force on the dipole $\boldsymbol{\mu} = (0, 0, m_0)$. For the case of a vertical dipole at a distance a above the superconductor one gets

$$\mathbf{F} = \frac{3\mu_0 m_0^2}{32\pi a^4}\hat{k} \tag{13.2}$$

Noteworthily, in order to avoid the extraneous appearance of a factor of 2, the gradient must be taken with care. In our case, considering that the distance between $\boldsymbol{\mu}$ and its image is $2a$ one should actually perform the operation $\partial/\partial(2a)$ if $\mathbf{B}$ is written in terms of equation (13.1). Physically, just consider that a virtual displacement $a \rightarrow a + \delta a$ implies an increment of $2\delta a$ in the distance between the dipoles.

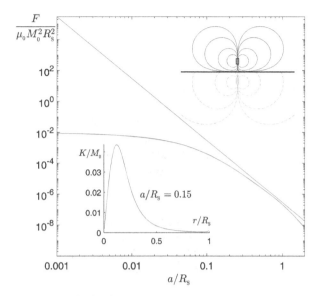

Figure 13.2. Comparison of the levitation force on a small magnet ($R_M = h_M = 0.1R_S$) as obtained from the dipole–dipole approximation (equation (13.3)) (blue) and from a more realistic model (equation (13.13d)) (red). Also shown is the sketch of magnetic flux lines and the 'image' dipole (upper inset). The lower inset shows the profile of induced sheet current K in the superconductor, for the separation distance $a = 0.15R_S$. Dimensionless units are used, with $M_0 \equiv m_0/(\pi R_M^2 h_M)$. 🔗. Source codes available at https://iopscience.iop.org/book/978-0-7503-2711-4.

Figure 13.2 has been created with the help of the following codes
- `plot_dipole_dipole.m` (main)
- `integ_F.m`
- `integ_K.m`
- `exact_cyl_g.m`
- `vec_pot_dip.m`
- `create_figure_dipole_dipole.m`

As explained within them, they include the expressions for the dipole–dipole model and a more realistic integral approximation developed later within this chapter.

Example

Equation (13.2) allows to obtain an estimation of what can be expected in a typical levitation experiment. Consider that one uses a conventional kit with a NdFeB magnet[2] ($\mu_0 M_0 = 1.17$ T) with dimensions $R_M = h_M = 2$ mm over a perfect super-conducting surface. Assuming a uniform magnetisation along the axis of the cylinder, the magnetic moment is given by $m_0 = M_0 \pi R_M^2 h_M$. Then, by equating the magnetic force to the weight of the magnet (a typical value for the density is $7.5\,\text{g cm}^{-3}$), one obtains a levitation height $a = 10.3$ mm, not far from what one may expect. In fact,

---

[2] Typically with YBCO pellets that may be operated using liquid nitrogen.

this somehow overestimates the real value, but being a rather crude calculation it works reasonably.

The above example will be used as a connecting thread throughout this chapter. Progressive improvements to the above calculation will be exposed below. By now, let us provide a means of delimiting the predictive power of the *dipole–dipole* approximation by comparing to a more refined model that allows to input both the actual size of the magnet and the superconductor. Thus, in figure 13.2 we check it against the (further explained) approximation that uses the actual size of the magnet and a finite radius for the superconductor. As expected, the rudimentary dipole–dipole model makes full sense in the region

$$R_M \ll a \ll R_S$$

Physically, this inequality means that the magnet should be close enough to the superconductor so as to have negligible effects due to supercurrents in the periphery ($a \ll R_S$), but also not that close ($R_M \ll a$) because for very small distances the point-like model fails due to the real size of the magnet. In fact, the small distance limit of equation (13.2): $F(a \to 0) \to \infty$ is clearly unphysical. In the following section, we provide a model that overrides this limitation by accounting for the finite size of the magnet. This will allow us to have a more realistic estimation of the maximum magnetic forces attainable.

Magnets of arbitrary size (image–magnet approximation)
Let us now consider the case described by the 'uncoupled conditions'

$$a \ll R_S$$
$$R_M \ll R_S$$

i.e. both the size of the magnet and the separation are small enough as compared to the size of the superconductor[3], but arbitrarily related one to the other. Then, as shown in [7], the repulsion force between the magnet and the superconductor may be evaluated by resorting again to the image technique.

For a cylindrical magnet uniformly magnetised along its axis, perpendicular to the surface of the superconductor, the image is just an identical magnet, symmetrically located and with antiparallel magnetisation. Then, by using the force equation between magnetic bodies one obtains

$$F(a) = 2\mu_0 M_0^2 R_M \left[2a_2 \frac{K(p_2) - E(p_2)}{p_2} \right.$$
$$\left. - a_1 \frac{K(p_1) - E(p_1)}{p_1} - a_3 \frac{K(p_3) - E(p_3)}{p_3} \right]$$

(13.3)

---

[3] This allows to treat the superconductor as an infinite sheet.

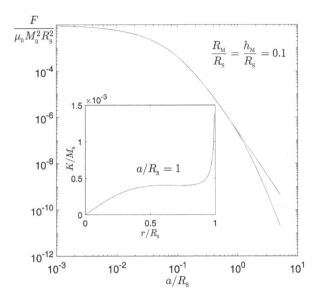

Figure 13.3. Comparison of the levitation force on a small magnet ($R_M = h_M = 0.1 R_S$) as obtained from the large superconductor approximation (equation (13.3)) (blue) and from a more realistic model (equation (13.13d)) (red). The inset shows the profile of induced surface current K in the superconductor for the separation distance $a = R_S$. Dimensionless units are used, with $M_0 \equiv m/(\pi R_S^2)$. $\mathscr{O}$. Source codes available at https://iopscience.iop.org/book/978-0-7503-2711-4.

with $p_n \equiv R_M/\sqrt{a_n^2 + R_M^2}$, $a_1 \equiv a$, $a_2 \equiv a + h_M/2$, $a_3 \equiv a + h_M$ and $K(p)$, $E(p)$ standing for the complete elliptic integrals of the first kind.

In order to establish the range of application of equation (13.3) we provide figure 13.3 and the related scripts. As shown in the plot, by contrast to the previous approximation the region of small distances is very well described. In fact, equation (13.3) is a useful instrument that allows to evaluate the 'maximum force' attainable between a given magnet and superconductor. It corresponds to the limit $F(a \to 0)$. On the other hand, also as expected, the approximation fails when the experiment reveals the finite size of the superconductor. Notice that for $a \gtrsim R_S$ the 'image' technique method gives an overestimated force as compared to more realistic models. Strictly speaking, the mirror image of the magnet gives the real physics of the problem if the superconductor may be considered an infinite sheet, and this picture fails as the magnet gets increasingly farther away.

Finally, recall that the profile of the supercurrent is no longer a bell shaped curve as for the case of small nearby magnets.

Figure 13.3 has been created with the help of the following codes
- `plot_semiinfinite.m` (main)
- `semiinfinite_sc.m`
- `integ_F.m`
- `integ_K.m`

- `exact_cyl_g.m`
- `create_figure_semiinfinite.m`

As explained within them, they include the expressions for the image–magnet model and the more realistic integral approximation developed later within this chapter.

Large distant magnets (uniform field approximation)
Another useful approximation, frequently mentioned in the literature is the so-called 'uniform field model'. Apparently, by uniform it is not meant that the field applied to the superconductor is strictly uniform, because this would lead to a vanishing force! Indeed, the idea is that the field produced by the magnet is basically constant over the surface of the superconductor but changes with vertical position, i.e. one has $B_0(a)$. As indicated in the forthcoming inequalities, this approximation makes sense when the size of the superconductor is small as compared to the magnet and to the distance between them:

$$a \gg R_S$$
$$R_M \gg R_S$$

Under such conditions, the superconductor basically shields a magnetic field that is well described by the expression of the field along the axis of a cylindrical magnet

$$B_0(a) = \frac{\mu_0 M_0}{2}\left[\frac{a + h_M}{\sqrt{(a + h_M)^2 + R_M^2}} - \frac{a}{\sqrt{a^2 + R_M^2}}\right] \tag{13.4}$$

As shown in [8], when such a field is shielded by a superconducting disk, a closed form expression may be found for the levitation force. Explicitly

$$F(a) = \frac{4 M_0 R_M^2 R_S^3}{3} B_0(a)\left\{\left[a^2 + R_M^2\right]^{-3/2} - \left[(a + h_M)^2 + R_M^2\right]^{-3/2}\right\} \tag{13.5}$$

Figure 13.4 illustrates that the 'uniform field' approximation works rather well for magnets of large size at non-small distances from the superconductor. By using the software associated to this figure, the reader may check that the approximation clearly fails as the magnet size becomes smaller (exercise R13-3). The following scripts are provided

- `plot_uniform.m` (main)
- `uniform_field.m`
- `integ_F.m`
- `integ_K.m`
- `exact_cyl_g.m`
- `create_figure_uniform.m`

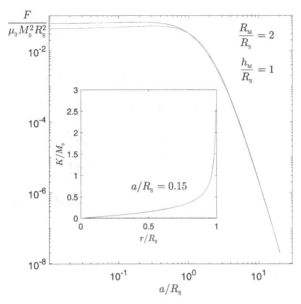

Figure 13.4. Comparison of the levitation force on a big magnet ($R_M = 2R_S$, $h_M = R_S$) as obtained from the uniform field approximation (equation (13.5)) (blue) and from the realistic model (equation (13.13d)) (red). The inset shows the profile of induced surface current K in the superconductor for the separation distance $a = R_S$. Dimensionless units are used, with $M_0 \equiv m_0/(\pi R_S^2 h_S)$. $\mathscr{O}$. Source codes available at https://iopscience. iop.org/book/978-0-7503-2711-4.

Cylindrical magnet above superconducting disk (exact solution)
Below, we develop a model that allows to dispense the approximations introduced in the above models. Although, in general, computer tools will be necessary for eventual evaluations, a number of closed form formulas will be provided and one can still name the model as 'analytical'. This will be possible with the only requirements of cylindrical symmetry and vanishingly small thickness for the superconductor. This exact model (any arbitrary relation between the quantities a, R_M, h_M, R_S is allowed) was presented in [8] and has been used as a realistic reference in figures 13.2, 13.3, and 13.4. Let us see how it arises.

Consider a superconducting disk with radius R_S embedded in some rotationally symmetric magnetic field around the same axis. We take the centre of the disk as the origin of coordinates. As conventionally, the cylindrical coordinates will be named after ρ, φ, z. The points of the disk will be defined by ($r \leqslant R_S$, $0 < \phi < 2\pi$, $z = 0$). The problem will be stated in terms of currents and vector potentials. Taking advantage of the problem's symmetry, let us divide the surface of the superconductor into a collection of circular loops, each carrying a current $dI = K(r)dr$.[4] Then, the response of the disk to the applied field may be written in terms of the vector potential as [6]

---

[4] Recall the definition of the sheet current $K(r) = \int J_\varphi(r, z)\, dz$.

$$A_{s\varphi}(\rho, z) = \frac{\mu_0}{4\pi} \int_0^{R_S} dr \int_0^{2\pi} d\phi \left[\frac{rK(r)\cos\phi}{\sqrt{r^2 + \rho^2 + z^2 - 2r\rho\cos\phi}} \right] \tag{13.6}$$

As shown in [8] this relation may be restated in the form

$$A_{s\varphi}(\rho, z) = \frac{\mu_0}{2} \int_0^\infty dk \left[A(R_s, k)J_1(kr)e^{-k|z|} \right] \tag{13.7}$$

with J_1 being the first-order Bessel function and the definition

$$A(R_s, k) \equiv \int_0^{R_S} dr \left[rK(r)J_1(kr) \right] \tag{13.8}$$

Notice that $A(R_s, k)$ plays the role of a 'Fourier–Bessel' expansion coefficient of the current density. In fact, as suggested in exercise R13-4 one may invert the above expression to obtain

$$K(r) = \int_0^\infty dk \left[kA(R_s, k)J_1(kr) \right] \tag{13.9}$$

Outstandingly, all the quantities of interest of our problem may be reconstructed by knowledge of $A(R_s, k)$. In order to solve for this function, recall that in terms of the above equations, the physical problem may be stated as

$$\frac{\mu_0}{2} \int_0^\infty dk \left[A(R_s, k)J_1(kr) \right] = -A_M(r, 0), \quad r \leqslant R_S \tag{13.10a}$$

$$\int_0^\infty dk \left[kA(R_s, k)J_1(kr) \right] = 0, \quad r > R_S \tag{13.10b}$$

Physically:
- Over the disk, one has a vanishing potential ($A_s + A_M = 0$). This ensures a vanishing magnetic flux density[5], and thus a perfect Meissner state.
- The current is confined to the surface of the disk.

The system of equations (13.10) is a non-trivial, but well known mathematical problem, widely used in some areas of physics such as fluid mechanics. It is an example of the so-called *dual integral equations*. One must solve for the unknown function $A(R_s, k)$ that depends on the parameter R_s. The reader may find an extensive guide to such problems in the book by I N Sneddon [9]. Skipping some technical details, it may be shown that the solution is given by

$$A(R_s, k) = \sqrt{\frac{2}{\pi}} \int_0^{R_S} dt[g(t)\sin(kt)] \tag{13.11}$$

with

$$g(t) = \sqrt{\frac{2}{\pi}} \int_0^t dr \frac{rb(r)}{\sqrt{t^2 - r^2}} \tag{13.12}$$

---

[5] Recall that $\oint \mathbf{A} \cdot d\ell = \Phi$.

For compactness, here we have used the notation $b(r) \equiv (2/\mu_0)B_{MZ}(r, 0)$ for the magnetic field created by the permanent magnet on the disk, whose position sets the origin of coordinates (i.e. $z = 0$ on surface).

Notice that, upon determining the magnetic source B_{MZ} is given and one may get the function $g(t)$, and from this $A(R_S, k)$, which is the key to deriving the physical quantities of interest (vector potential of the superconductor, induced currents, etc). Outstandingly, the response of the superconductor may be obtained based upon of the normal component of the applied magnetic fields at the surface[6]. In conclusion, upon obtaining the corresponding $g(t)$, one may analyse the effect of different magnetic sources on the induced currents, arising forces, and so on. Thus, after some algebra with electromagnetic relations, one may obtain[7]

$$\text{Induced current} \qquad K(r) = -\sqrt{\frac{2}{\pi}} \frac{\partial}{\partial r} \int_r^{R_S} dt \frac{g(t)}{\sqrt{t^2 - r^2}} \qquad (13.13a)$$

$$\text{Induced field} \qquad B_{SZ}(0, z) = \sqrt{\frac{2}{\pi}} |z| \int_r^{R_S} dt \frac{tg(t)}{t^2 + r^2} \qquad (13.13b)$$

$$\text{Self - energy} \qquad U_s = \frac{\mu_0 \pi}{2} \int_0^{R_S} dt[g(t)]^2 \qquad (13.13c)$$

$$\text{Repulsion Force} \qquad F(a) = -\frac{\mu_0 \pi}{2} \frac{\partial}{\partial a} \int_0^{R_S} dt[g(t)]^2 \qquad (13.13d)$$

From the practical side, in general, the application of the above expressions requires the numerical integration of some given functions. Thus, starting with the actual field of interest for a given situation $B_{MZ}(r, 0)$ one must perform the integration in (13.12) so as to obtain $g(t)$.[8] Then, with $g(t)$ at hand, one must integrate again so as to obtain any of the above displayed physical quantities. In order to see how this works, the reader may just review the software provided for the previous figures, in which several selections of the magnetic source $B_{MZ}(r, 0)$ have been used. For instance, the MATLAB function integ_K.m has been used to obtain the profile $K(r)$ by means of expression (13.13a).

13.1.2 Meissner state: numerical solution

Still keeping within the approximation of rotational symmetry, in this section we provide the resources that allow to analyse the levitation force properties in the

---

[6] A fact that, on the other side, could be advanced following intuition because flux expulsion consists of counterbalancing such component, actually.

[7] In the set of equations (13.13) one must keep in mind that, though not explicit, the function $g(t)$ will be dependent on the distance between the magnet and the superconductor a. Such dependence appears when the actual expression of the magnetic field over the superconductor is introduced in equation (13.12).

[8] Occasionally, $g(t)$ admits a closed form expression as for the case of a dipole or cylindrical permanent magnet (see exercises R13-2 and R13-5).

general situation depicted in figure 13.1, i.e. a cylindrical magnet of arbitrary size close to the superconducting cylinder of arbitrary size.

The numerical statement of the levitation force problem in the Meissner state will be built upon the expression of the energy given in equation (2.9). The configuration of induced supercurrents may be found by minimising such expression, a rather simple task with the help of MATLAB. With such distribution at hand, magnetic forces will be also obtained straightforwardly by accumulation of forces between individual current loops, which will be the elements of the discretized statement. More specifically, the repulsion force between the superconductor and the magnet is nothing but the summation of forces between each supercurrent loop and each equivalent magnetisation loop in the magnet[9].

Let us see how this arises[10]. Recalling the electromagnetic field manipulations suggested in exercise R1-2, equation (2.9) may be cast in the equivalent, but more convenient form for our purposes

$$U[\mathbf{J}(\mathbf{r})] = \frac{\mu_0}{8\pi} \int \int_\Omega \frac{\mathbf{J}(\mathbf{r}) \cdot \mathbf{J}(\mathbf{r}')}{\|\mathbf{r} - \mathbf{r}'\|} dV dV' + \frac{\mu_0 \lambda_L^2}{2} \int_\Omega J(\mathbf{r})^2 dV$$
$$+ \int_\Omega \mathbf{A}_M(\mathbf{r}) \cdot \mathbf{J}(\mathbf{r}) \, dV + \text{constant}$$

(13.14)

Here, as before, Ω stands for the region occupied by the superconductor, $\mathbf{J}(\mathbf{r})$ for the supercurrent distribution, $\mathbf{A}_M(\mathbf{r})$ for the magnetic potential created by the magnet, and a constant term indicates the self-energy of the permanent magnet, that is assumed to remain constant (the magnet is not 'demagnetised' by the interaction). Additionally, we have used that the interaction between the magnet and the superconductor may be written as a single term due to symmetry (i.e. $U_{SM} + U_{MS} = 2U_{SM}$).

Now, focusing on the discretized formulation of the problem, we introduce the notation introduced in chapter 8. In matrix form, the energy of the system is reexpressed as

$$\mathbf{U} = \frac{1}{2}\langle \mathbf{J}|\mathbf{m}|\mathbf{J}\rangle + \langle \boldsymbol{\psi}_{Sm}|\mathbf{J}\rangle + \frac{1}{2}\langle \mathbf{J}|\mathbf{\Delta}|\mathbf{J}\rangle$$

(13.15)

with $\mathbf{m}$ and $\boldsymbol{\psi}_{Sm}$ standing for the inductance matrix and the magnetic potential as already defined in chapter 8. $\mathbf{\Delta}$ stands for the diagonal matrix that couples each circuit with itself, so as to account for the kinetic term which includes the London penetration depth[11].

---

[9] Recall that, physically the magnet is equivalent to a system of surface currents (equation (8.31)).

[10] The procedure generalises the ideas introduced in section 9.2 for the perfect Meissner state under uniform applied field.

[11] Although in our application this matrix will be proportional to the identity, in general, it could be used for encoding problems with spatial inhomogeneities or anisotropy (this also applies to equation (13.14), where one could use the more general expression $\mathbf{J}(\mathbf{r}) \cdot [\mathbf{\Delta}(\mathbf{r})\mathbf{J}(\mathbf{r})]$, with $\mathbf{\Delta}(\mathbf{r})$ a position dependent tensor with unequal diagonal elements.

Following the numerical scheme introduced in chapter 8, the supercurrent distribution induced by a given magnetic source ψ_{Sm} may be obtained by minimising U as a function of the current density in the finite elements $\langle \mathbf{J} | \equiv (J_1, J_2, \dots, J_n)$ (see equation (8.3)), i.e. collection of elementary circuits for the problem. Apparently, for the current situation, these circuits will be a grid of circular loops covering the volume of the superconductor.

From the practical point of view, and keeping in mind that here we are focused on the evaluation of forces in macroscopic systems, the approximation $\lambda_L \to 0$ allows a further simplification. We will proceed by considering that the supercurrents flow in a very thin layer, in an essentially hollow medium, and use the principle

$$\text{minimise } \mathbf{U} = \frac{1}{2}\langle \mathbf{K}|\mathbf{m}|\mathbf{K}\rangle + \langle \psi_{Sm}|\mathbf{K}\rangle \qquad (13.16)$$

with K being the notation for the sheet current, as before. Here, $|\mathbf{K}\rangle$ it will be interpreted as a 'vector' whose components are the values of the sheet current at the collection of imaginary circuits covering the surface of the superconductor: $\langle \mathbf{K}| \equiv (k_1, k_2, \dots, k_n)$

Mathematically, as the minimisation of $\mathbf{U}$ may be performed without constraints, and the function to be minimised is quadratic in the variables, a straightforward method may be used based on matrix inversion, i.e. the use of numeric minimisers is not required. Recall that the minimum is just given by

$$|\mathbf{K}\rangle_{min} = -\mathbf{m}^{-1}|\psi_{Sm}\rangle \qquad (13.17)$$

with $\mathbf{m}^{-1}$ being the inverse of the inductance matrix for the superconducting elements and $|\psi_{Sm}\rangle$ the vector that gives the potential created by the magnet at the positions of the superconducting elements in column vector form. Interestingly, this simplified model is supported by a straightforward interpretation and contains the unabridged physics of the problem. Just recall that, macroscopically, the role of λ_L is basically to establish the decay length for the fields. Within the approximation $\lambda_L \to 0$, one may consistently assume the circulation of current in a thin layer, by confining the finite element circuits $1, 2, \dots, n$ to a collection of points at the cylinder's surface (see figure 13.5). Then, the numerical results will be physically meaningful if the real dimensions of the system are much bigger than λ_L.[12]

On the other hand, with help of the software in figure 13.5 one may check that the current density distribution that is obtained within the thin disk limit may be considered a good approximation in many instances. As shown, for the case of small magnets, the numerical solution indicates that screening currents are typically

---

[12] For the investigation of mesoscopic systems, one could in fact use the same procedure by replacing $\mathbf{m} \to \mathbf{m} + \mathbf{\Delta}$ in (13.17) and using the full 2D grid covering the volume of the superconducting cylinder (exercise C13-2).

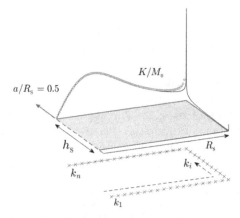

Figure 13.5. Cross section plot representing the surface current distribution induced in a superconducting cylinder ($R_S = 1.4h_S$) by the field of a nearby cylindrical permanent magnet (see figure 13.1). The superconductor is described by a set of surface current circuits (k_i), along the perimeter. Each of them stands for an elementary circular loop. The magnet, of dimensions $R_M = h_M = 0.1R_S$ settles at $a = 0.5R_S$. On top of K(r) we plot (red symbols) the solution from the thin disk approximation (equation (13.13)). 🔗. Source codes available at https://iopscience.iop.org/book/978-0-7503-2711-4.

concentrated on the upper (closer to magnet) surface of the superconductor, with the profile rather close to the prediction from such model.

Figure 13.5 has been created with the help of the following codes
- `cyl_meiss_lev.m` (main)
- `generate_grid_surf.m`
- `matrixM.m, mutual.m, self.m`
- `psi_pot_magnet.m`
- `integ_K.m, exact_cyl_g.m`
- `plot_meiss_num.m, create_figure_meiss_num.m`

Calculation of the repulsion force
As said above, having obtained the supercurrent distribution, and using the equivalence of the magnet to a distribution of surface currents, deriving the levitation force may be stated as a summation problem. Alternatively, by using a 'virtual displacement' argument this may be implemented as follows[13].

We start with the interaction term[14] in equation (13.15), from which the normalised force ($f \equiv F/\mu_0 M_0^2 R_S^2$) is

$$\mathbf{f} = \mathbf{grad}_2 \langle \psi_{Sm} | \mathbf{K} \rangle = \langle \mathbf{grad}_2 \, \psi_{Sm} | \mathbf{K} \rangle \qquad (13.18)$$

where $\langle \mathbf{grad}_2 \psi_{Sm} |$ is a finite element row vector evaluated at the position of the superconducting elements, and ψ_{Sm} and $\mathbf{K}$ are expressed in units normalised by $(\mu_0 M_0 R_S^2)$ and (M_0) respectively. As usual, this operation is conveniently implemented

---

[13] The equivalence of both methods is studied in exercise R13-7.
[14] Analysing the self-energy, on the other hand, may be used to evaluate internal stresses.

in terms of mutual inductance matrices. Thus, if one uses $m_{\alpha i}$ for the element coupling the αth equivalent magnetisation current circuit and the ith superconducting circuit,

$$\mathbf{grad}_2 \, \psi_{\mathrm{Sm},i} = \sum_\alpha \mathbf{grad}_2 \, m_{\alpha i} \equiv \sum_\alpha \mathbf{n}_{\alpha i}. \qquad (13.19)$$

Apparently, $\mathbf{n}_{\alpha i}$ is a geometric coefficient that may be determined *a priori* based on the symmetry of the setup and of the displacement. Assuming the arrangement in figure 13.1 (circular surface current loops) and a shift along the z-axis one has

$$f_z = \sum_{i,\alpha} n_{\alpha i}^z \, k_i \qquad (13.20)$$

with

$$n_{\alpha i}^z = \frac{z_\alpha - z_i}{\sqrt{(r_\alpha + r_i)^2 + (z_\alpha - z_i)^2}} \left[\mathrm{K}(p_i) - \frac{r_\alpha^2 + r_i^2 + (z_\alpha - z_i)^2}{(r_\alpha - r_i)^2 + (z_\alpha - z_i)^2} \mathrm{E}(p_i) \right] \qquad (13.21)$$

Here, we have used the definition

$$p_i = \sqrt{\frac{4 r_\alpha \, r_i}{(r_\alpha + r_i)^2 + (z_\alpha - z_i)^2}}$$

and resorted to the well known expression for the mutual inductance between two coaxial circular loops of current (8.20).

Figure 13.6 illustrates the application of equation (13.20). In brief, the levitation force for a given arrangement is obtained by (i) calculating the supercurrent distribution as a function of distance, i.e. $\mathbf{K}(a)$, and then (ii) performing the summation in equation (13.20) for each distance. We show the results obtained for two different cases ('small' and 'big' magnets). The plot includes the comparison to the analytical results under the thin disk approximation. Notice that, for the case of small magnets, the disk approximation performs rather well for the whole range of distances. In contrast, if the magnet's diameter is larger that the superconductor's important differences appear. As the reader may check by means of the accompanying software, for the range $R_M > R_S$ the amount of current that flows in the lateral and lower surfaces of the superconducting cylinder is non-negligible, and one must use the numerical 2D model for the Meissner current if a realistic estimation of the levitation force is to be obtained.

Figure 13.6 has been created with the help of the following codes

- `cyl_meiss_lev2.m` (main)
- `generate_grid_surf.m`
- `matrixM.m`, `mutual.m`, `self.m`
- `psi_pot_magnet.m`
- `force_profile_meissner.m`, `force_magnet.m`, `force_2_loops.m`
- `integ_F.m`, `exact_cyl_g_2.m`
- `plot_force_meiss_num_disk.m`, `create_figure_f_num_disk.m`

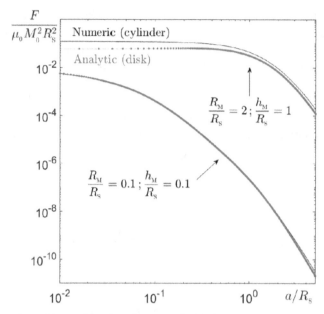

Figure 13.6. Comparison of the repulsion force between cylindrical magnets and superconductors (sketched in figure 13.1) as obtained from the numerical 2D model in equation (13.20) (black lines) and from the analytical 1D thin disk model (equation (13.13d)) (red symbols). Two different conditions ('small' and 'big' magnets) are displayed. 🔗. Source codes available at https://iopscience.iop.org/book/978-0-7503-2711-4.

13.2 Levitation in the critical state

Our next step in building increasingly realistic models that have an application in technological systems will be to consider the critical state approximation (recall 3.3) for the superconducting material. On the one side, in order to enable realistic working conditions, i.e. high enough temperatures (with liquid nitrogen's as a reference) and high enough magnetic fields (in the range from 1 to several Tesla) one is lead to use extreme type-II superconductors (typically YBCO, and MgB_2 as a promising candidate at lower temperatures). On the other side, the operational conditions in real-life MagLev systems will be such that one needs a fundamental modification of the above introduced Meissner state modelling. As shown below, the sequence of events that define the configuration of the magnet/superconductor arrangement, which is of no relevance in the former limit will be, as a matter of fact, an essential mechanism in superconductors with pinning. Hysteresis effects will come on the scene.

13.2.1 The issue of mechanical stability

Resorting to figure 8.7 one may believe that a type-II superconductor with pinning expels the magnetic flux of a permanent magnet in a similar fashion to the above considered Meissner limit. In fact, the higher the value of the critical current density J_c, the thinner the shell of circulating currents that are needed to create a field opposing to the magnet's. More specifically, for the cylindrical symmetry, the thickness of such a layer may be estimated by comparison to figure 13.4. One has $J_c\,d \sim K \lesssim 3M_0 \Rightarrow d \lesssim 3M_0/J_c$. According to this[15], one could infer that

- the 'Meissner state' approximation provides the limiting value of the levitation force for for type-II materials with elevated critical currents,
- the behaviour of the high-J_c materials may be roughly deduced by studying the Meissner limit for a given geometry.

The first statement will happen to be fully valid, though one has to be careful with the second one. In fact, as it will be shown later, there is a fundamental issue that distinguishes both situations. Contrary to what occurs in the Meissner state, the force between the magnet and superconductor strongly depends on *history effects* for type-II materials. Interestingly, this property is not always a handicap, as it connects to an essential mechanism that allows to provide *stability* in real levitation systems.

In physical terms, as it has been well known for nearly two centuries [10], equilibrium through magnetic forces is inherently unstable. Thus, based on everyday's life experience with permanent magnets[16], one could guess that a magnet would not float so easily above a superconductor. In fact, by recalling the dipole–dipole model of the previous section, one can already say that a lateral restoring force around equilibrium does not seem to be granted. At most, one may expect that for a tiny magnet above a superconductor, small lateral displacements around the central equilibrium point produce *neutral* equilibrium because supercurrents rearrange and the image–magnet re-allocates. Real facts are even less attractive as one may check by applying the numerical method introduced in section 13.1.3. Thus as shown in figure 13.7, a small magnet above a flat superconductor is basically at neutral equilibrium for central positions of the magnet, while it experiments instability for lateral displacements. Concerning the actual application of equation (13.17) for the geometry of figure 13.7 with current circuits in the form of long filaments, one must recall that, physically induced currents should form loops. In order to enforce this condition one should add the constraint $\sum_i k_i = 0$. As indicated within the accompanying software, the constraint maybe implemented by the use of a Langrange multiplier, which produces an *augmented* version of the linear system, i.e.

$$|\mathbf{K}\rangle_{\text{aug}} = -\mathbf{m}_{\text{aug}}^{-1}|\psi_{\text{Sm}}^{\text{aug}}\rangle \qquad (13.22)$$

Here, $\mathbf{m}_{\text{aug}}$ is the matrix

$$\mathbf{m}_{\text{aug}} = \begin{bmatrix} m_{11} & m_{12} & \cdots & m_{1n} & -1 \\ m_{21} & m_{22} & \cdots & m_{2n} & -1 \\ \vdots & & \ddots & \vdots & \vdots \\ m_{n1} & m_{n2} & \cdots & m_{nn} & -1 \\ -1 & -1 & \cdots & -1 & 0 \end{bmatrix}$$

---

[15] In this chapter we will provide the tools for the quantitative analysis of these aspects.

[16] As the reader may experiment with, it is not possible to achieve levitation with two magnets, neither flotation by opposing equivalent poles, nor suspension by opposing contraries.

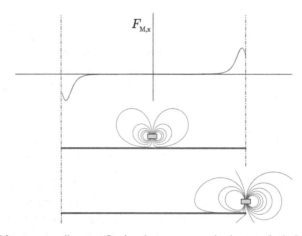

Figure 13.7. Lateral force on a small magnet floating above a superconducting tape (in the Meissner state) in terms of the lateral position of the magnet (at given height). Below we plot the magnetic field structure for two given positions of the magnet: right above the centre of the tape and just above the edge. The tape is infinitely long across the plane of the picture. 🔗. Source codes available at https://iopscience.iop.org/book/978-0-7503-2711-4.

with the mutual inductance (per unit length) elements

$$m_{ii} = \frac{\mu_0}{8\pi}$$

$$m_{ij} = \frac{\mu_0}{4\pi} \ln \frac{r^2}{(x_i - x_j)^2 + (z_i - z_j)^2}$$

(13.23)

that correspond to parallel long wires at normalised positions (x_i, z_i) and (x_j, z_j) and stretching along the y-axis[17].

On the other hand, and with the aim of keeping calculations at the 2D level, we have also considered a long magnet parallel to the superconductor and uniformly magnetised along the vertical axis. Then, magnetisation surface currents may also be taken in the form of long wires extending across the vertical sides of the magnet and forces calculated by summation of the expression[18]

$$\frac{\mathbf{F}_{\alpha j}}{\mu_0 M_0 J_c W^2} = \frac{J_j}{2\pi} \frac{(x_j - x_\alpha, 0, z_j - z_\alpha)}{(x_j - x_\alpha)^2 + (z_j - z_\alpha)^2}$$

(13.24)

that gives the interaction between the equivalent α-wire of magnetisation and the jth superconducting wire, which carries a normalised current density J_j.

---

[17] These expressions have been obtained by application of the definition $M_{ij} = \Phi_{ij}/I_j$, and assuming that, in general one may neglect the radius r of the wires when compared to their distances. For a thorough study about mutual inductances the reader may see [11] (in Gaussian units).

[18] Alternatively, one may also proceed by evaluation of the interaction energy between the magnet and the superconductor and taking the numeric derivative. In exercise R13-7 we suggest to check one method against the other.

Owing to the finite size of the superconductor, the lateral force (x -component) on the magnet in terms of the lateral position, although nearly zero close to the centre, shows a clearly unstable behaviour. Any however small displacement of the magnet inevitably leads to slippage and downfall along the side. It is just for this reason that Arkadiev [5] used a bowl shaped superconductor. As a matter of fact, curvature modifies the related interaction and produces stability (restoring force) by modifying the profile displayed in figure 13.7. The reader may show it by solving exercise C13-1.

Considering the previous analysis, the question arises: why is then relatively simple to achieve stable levitation of magnets above flat high-T_c type-II super-conductors? The reason is that, actually, the underlying physical mechanism is rather different. Thus, in type-II superconductors cooled with a nearby magnet, flux dynamics is not determined by the formation of thermodynamic equilibrium states. In contrast, it takes place in the form of highly dissipative processes that may produce (meta)stable configurations related to such conditions. The rather complex underlying flux dynamics may be reasonably described through the *minimum entropy production* principle introduced in 1.2.2 and extensively used in chapter 8, for instance, as the fundamental physics behind the critical state approximation. In particular, if one applies the method developed in section 8.4, it becomes apparent that a stable situation may be achieved (see figure 13.8). As a first step towards the complex scenario of levitation systems with type-II materials, let us discuss the modification of the results for the geometry considered in figure 13.7 by incorporating the flux pinning mechanism.

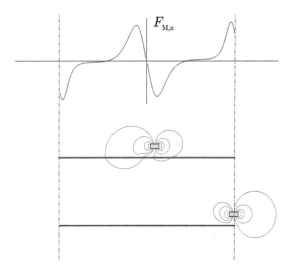

Figure 13.8. Lateral force on a small magnet above a type-II superconducting tape (in the *critical state*) in terms of the lateral position. The superconductor was cooled down with the magnet centred, near the surface. Below we plot the magnetic field lines for two given positions of the magnet: close to the initial position above the centre of the tape, and just above the edge of the superconductor. 🔗. Source codes available at https://iopscience.iop.org/book/978-0-7503-2711-4.

One may notice that a lateral restoring force (guidance) arises. Thus, $F_{M,x}$ opposes to the magnet's displacement (negative force for positive drift and vice versa) within a noticeable range of distances. The underlying physical mechanism reveals in the calculated flux line structure. One may relate the rigidity of the system to the 'trapping' of flux lines within the superconductor. For moderate shifts of the magnet, the superconductor reacts so as to avoid changes in the profile of magnetic field that was present in the cooling process. Also, as expected, the plot shows that stability is lost for larger displacements.

Technically, figure 13.8 has been created through the step-by-step resolution of equation (8.34) with the initial conditions: $|\mathbf{J}\rangle = |0\rangle$ (null supercurrents) and $|\psi^{Sm}\rangle$ given by the nearby magnet's vector potential. This means that the starting magnetic profile within the superconductor is taken as the frozen magnet's profile. It is the critical state hypothesis itself (opposition to flux variations through a highly dissipative mechanism of flux pinning) that explains the actual response of the system.

Figures 13.7 and 13.8 have been created with the help of the following codes:

- `tape_meiss_lev.m` (main for 13.7)
- `tape_cs_lev.m` (main for 13.8)
- `generate_grid_tape.m`
- `matrixM_tape.m`, `mutual_tape.m`
- `psi_pot_tape.m`, `psi_pot_tape_SC.m`
- `plot_meiss_tape.m`, `create_figure_meiss_tape.m` (for 13.7)
- `plot_cs_tape.m`, `create_figure_cs_tape.m` (for 13.8)

13.2.2 Critical state: analytical solution

As said before, generally speaking, the inclusion of pinning phenomena within superconductors in levitation systems drives the problem numerically. Nevertheless, a smart interpretation of the physics of extremely hard superconductors, i.e. with very strong pinning forces, allows to perform some analytical calculations that give reliable results in some conditions. Below, we briefly review the so-called 'frozen flux' model issued by A Kordyuk in 1997 [12]. Essentially, the model may be named as an *advanced mirror image method*, that upgrades the dipole–dipole model introduced in section 13.1.1 for the Meissner state limit.

Recall that the underlying concept that justified the original method was that for small magnets, not too close to the (flat) superconductor, one may implement the image technique in magnetostatics and replace the supercurrents by an image dipole whose action allows to reproduce the boundary conditions, i.e. perfect flux expulsion. Such a model may be applied to systems containing hard superconductors only under special circumstances. If the superconductor is cooled far apart from the magnet, pinning forces will only have the effect of reducing flux penetration to a small surface layer and equation (13.2) will reasonably describe the facts. Although the underlying physics is different[19], for such conditions the response of the system is

---

[19] An extreme hard superconductor basically acts as a *perfect conductor* that expels flux variations.

basically indistinguishable. However, if one has volume flux pinning effects induced by cooling the superconductor close to the magnet, the analogy fails.

Within the advanced image model, the critical state physics in hard superconductors is captured as follows. For a small permanent magnet of moment $\boldsymbol{\mu}_0$, initial position $\mathbf{r}_0$ and subsequent orientation $\boldsymbol{\mu}$ and position $\mathbf{r}$, one introduces:

- A 'frozen' image dipole that mimics the action of the supercurrents that are induced so as to retain the initial trapped field within the field cooling process. Such a dipole, of value $-\boldsymbol{\mu}_0^*$ lies at position $\mathbf{r}_0^*$, with the operation '*' giving the mirror reflection of the given vector by the surface of the superconductor.
- An 'active' image dipole that is updated along the subsequent movement of the magnet and standing for the dynamically induced currents due to Faraday's law. This dipole takes the value $\boldsymbol{\mu}^*$ and for each step sits at the position $\mathbf{r}^*$.

Let us illustrate how the model applies, by considering a simple example based on the arrangement that produced equation (13.2) for the Meissner state regime. Thus, taking $\mathbf{r}_0 = a_0\hat{k}$, $\boldsymbol{\mu}_0 = -m_0\hat{k}$, $\mathbf{r} = a\hat{k}$, $\boldsymbol{\mu} = -m_0\hat{k}$ and replacing values in equation (13.1), one may calculate the magnetic field produced by the image dipoles $\boldsymbol{\mu}_0^*$ and $\boldsymbol{\mu}^*$ at the position of the magnet

$$\mathbf{B}_{\text{images}} = \mathbf{B}_{\text{frozen}} + \mathbf{B}_{\text{active}} \tag{13.25}$$

Then by using $\mathbf{F} = (\boldsymbol{\mu}\nabla)\mathbf{B}_{\text{images}}$ one gets the force on the magnet[20]

$$\mathbf{F} = \frac{3\mu_0 m_0^2}{2\pi}\left[\frac{1}{16a^4} - \frac{1}{(a + a_0)^4}\right]\hat{k} \tag{13.26}$$

It is worth commenting that this expression asymptotically goes to equation (13.2) as a_0 increases. This is nothing but the zero field cooling limit, that as said is well represented by the Meissner state equations for superconductors with high critical currents. Also of interest is to notice that, for the initial steps of the magnet, i.e. $a \approx a_0$ one has $F \approx 0$. In fact, the image dipoles tend to cancel each other. Physically, the superconductor is penetrated by the magnet's trapped flux lines, but macroscopic currents do not still exist $\nabla \times \mathbf{B}_{\text{magnet}} = 0$ within the superconductor.

Recall that with equation (13.26) at hand, one may obtain information about the equilibrium properties of levitation systems with hard superconductors: stability, oscillation frequency, non-linear behaviour and other, in terms of the cooling distance and orientation. Nevertheless an important shortcoming remains. One has to upgrade the model so as to include hysteresis effects.

---

[20] As said when obtaining equation (13.2), one must take care when evaluating the gradient of the magnetic field created by the induced dipoles. Here the rule $\partial/\partial(2a)$ applies to the active image, that changes position with the original magnet.

13.2.3 Critical state: numerical solutions

Below, we will upgrade the resources provided with figure 13.8 so as to include physical modelling of some relevant properties in superconducting levitation systems. Based on 2D simulations, we bring tools for investigating hysteretic force-displacement curves. Related issues as the mechanical stability against arbitrary displacements of the magnet will be discussed.

Hysteresis in the (vertical) levitation force

A well known property of practical levitation systems is that force-distance measurements display a more or less pronounced hysteretic behaviour, i.e. for cyclic variations of the relative position magnet/superconductor one records different values of the force for the same distance. The phenomenon was described in section 4.4 and is modelled here. Figure 13.9 displays the levitation force on a cylindrical magnet that is brought close to a superconductor and then raised again. As in the previous paragraph, simulations of the superconducting response have been performed by using the software developed in section 8.4. A mesh of circular circuits covering the section of the superconductor has been used, and forces evaluated according to equation (13.20). Standard experimental parameters [13] have been considered for the simulation:

- Permanent magnet: NdFeB cylinder of radius 22.5 mm, height 15 mm and axial magnetisation $\mu_0 M_0 = 1.17$ T.
- Superconductor: YBCO cylinder of radius 25 mm, height 15 mm and critical current density $J_c = 10^8$ A/m^2.

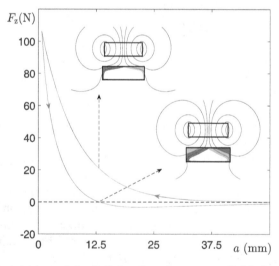

Figure 13.9. Hysteretic behaviour of the levitation force between a cylindrical permanent magnet and superconductor (standard experimental parameters as described in the text). The upper branch corresponds to the descending route of the magnet ($a = 50 \rightarrow 0.7$ mm), and the lower branch to the way back. The magnetic field and penetration profile of the current (as in figure 8.7) are shown for the distance $a = 12.5$ mm, at which the vertical force becomes zero in the ascending branch. $\mathscr{O}$. Source codes available at https://iopscience.iop.org/book/978-0-7503-2711-4.

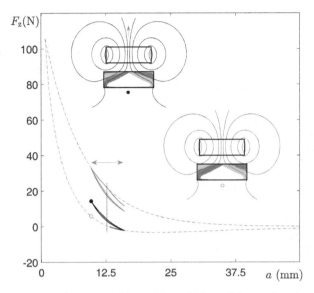

Figure 13.10. For the same experimental conditions of figure 13.9: small hysteresis cycles around the distance $a = 12.5$ mm induced by oscillations in the vertical position of the magnet. The magnetic profile for the selected distance $a = 9.3$ mm previous and subsequent to the oscillations is shown. $\mathscr{O}$. Source codes available at https://iopscience.iop.org/book/978-0-7503-2711-4.

As shown in [13], the critical state modelling for the superconductor together with the ansatz of uniform magnetisation of the magnet provide a very reasonable simulation of the experimental facts.

Concerning the observation of hysteresis in the levitation force, the insets of figure 13.9 provide the physical explanation of this phenomenon. The critical current profiles induced by moving the magnet 'penetrate' into the superconductor from top to bottom. The direction of circulation of the current (clock- or anticlock-wise) is determined by the actual shift of the magnet through Faraday's law. Owing to the flux pinning phenomenon, the profiles freeze when the variations cease. Thus, the actual configuration may be rather different for the same final conditions. As shown in the plot, this means for instance that for the separation $a = 12.5$ mm, one may either have $F_z = 21.6$ N or $F_z = 0$, depending on the branch of the process (a: 50 → 0 → 50 mm).

Also of technical interest is the consideration of small hysteresis cycles. Thus, for a certain system, one may have to describe what occurs if a small perturbation takes place around the equilibrium height at which the system levitates. Apparently, for small vertical perturbations, the magnet will begin to oscillate[21]. Also, a shift in the levitation point may occur owing to the re-establishment of the superconducting currents. The situation is illustrated in figure 13.10. Through the related software, one may evaluate the evolution of the vertical force after a vertical perturbation around some given point. Small hysteresis cycles may be observed for a given 'master' curve

[21] One must just consider the competition between gravity and the magnetic levitation forces.

F_z, that was determined by the cooling process of the system. As the levitation height is determined by intersecting the curve with the horizontal $F = mg$, small re-allocations are explained by the *transfer* of the magnet to a new position of equilibrium.

Figures 13.9 and 13.10 have been created with the help of the following codes

- `cyl_cs_lev_qp.m` (main for 13.9)
- `cyl_cs_lev_qp_small.m` (main for 13.10)
- `generate_grid.m`
- `matrixM.m, mutual.m, self.m`
- `psi_pot_magnet.m, psi_pot_SC.m`
- `force_profile_critical_state.m`
- `force_profile_critical_state_upper.m,`
- `force_profile_critical_state_lower.m` (for 13.10)
- `force_magnet.m, force_2_loops.m`
- `build_figure_lev_cycle.m` (for 13.9)
- `build_figure_lev_cycles.m` (for 13.10)
- `subfigure_desc.m, subfigure_asc.m, subfigure_asc_cycle.m`

Hysteresis in the (horizontal) guidance force

As said above, disturbances around the equilibrium point of levitation are an essential topic in practical levitation systems. As seen, vertical disturbances lead to small cycles within the $F_z(a)$ dependence. On the other hand, the reader may be wondering about the relevance of horizontal perturbations. In fact, recalling figure 13.8 one may suspect that also horizontal displacements may lead to restoring forces and lead to new equilibrium points.

In engineering terms, one speaks about 'guidance' forces in the following sense: if the magnet shifts to the right $x \to x + \delta x$, such an arising force will push it back to the left (say $F_x = -k\delta x$ with $k > 0$). In order to quantify these aspects, we put forward figure 13.11 and the related software. Here, aiming at the enabling 2D simulations, guidance forces will be analysed for translationally symmetric systems[22]. Thus, the magnet/superconductor system displayed in figure 13.11 must be interpreted as the cross section of a long permanent magnet, just on top of a superconductor both having a rectangular cross section, and with parallel long dimension (perpendicular to the plot). The magnet is uniformly magnetised along the vertical direction. Arbitrary values of the cross section dimensions have been chosen, and all distances normalised by the superconductor's half-width W. In order to obtain realistic simulations, the same material parameters as in the previous section are used. As is customary in systems with translational symmetry, here we obtain the force per unit length.

From the physical point of view, the *long rod* approximation used here is in fact a rather good representation of the *levitation train track geometry*, which consists of a long permanent magnet track and a finite, though long superconducting bars within the vehicle on top.

---

[22] Recall that, strictly speaking, for real systems, a lateral disturbance from the equilibrium position generates a 3D problem. Nevertheless the 2D approximations is a very reasonable description for transportation systems.

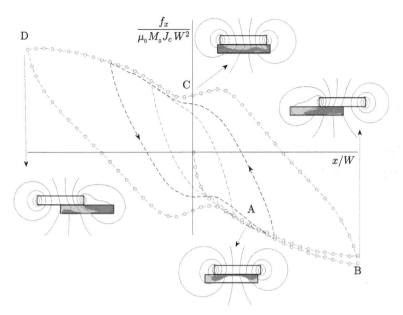

Figure 13.11. Horizontal (guidance) force on a permanent magnet that is moved periodically along the sequence $0 \to A \to B \to C \to D \cdots \circlearrowright$ on top of a superconductor. Both have the shape of long bars of rectangular cross section. Hysteresis cycles are shown for three amplitudes of oscillation ($0.25W$, $0.5W$, W), with W being the half-width of the superconductor. f_x is the force per unit length. $\mathscr{O}$. Source codes available at https://iopscience.iop.org/book/978-0-7503-2711-4.

As concerns the technical details for solving the critical state problem (section 8.4), here we use the same method employed in figure 13.8, but now related to a mesh of wires all across the rectangular section of the superconductor. Then, mutual inductances are evaluated through equation (13.23) and forces through equation (13.24).

The basic phenomenology that was presented in figure 13.8 is now upgraded by including the finite thickness of the superconductor, and a magnet of comparable dimensions. Recall that when forth and back lateral displacements occur a hysteretic response is induced, that is clearly dependent on the amplitude of such oscillations. Also, in this case, long range restoring forces are obtained, as well as an apparently shifted equilibrium position ($F_x = 0$), that strongly depends on the amplitude of the disturbance.

Remarkably, the representation of the magnetic field lines offers a rather intuitive picture of the underlying interaction. As said, the pinning of flux lines in the superconductor acts so as to retain the frozen magnetic flux profile. By shifting the permanent magnet, supercurrents are induced that produce a magnetic field that tends to compensate variations within the sample. This is especially visible for the snapshots at the positions $x/W = \pm 1$.

Figure 13.11 have been created with the help of the following codes

- `bar_cs_lev_qp_octave.m` (main for users of OCTAVE)
- `bar_cs_lev_fmincon_matlab.m` (main for users of MATLAB)
- `generate_grid_bar.m`

- matrixM_tape.m, mutual_tape.m
- psi_pot_tape.m, psi_pot_tape_SC.m
- cost.m (to be used in the main script of MATLAB)
- build_figure_guidance.m
- small_cycle.m, medium_cycle.m, large_cycle.m
- subfigure_decrease.m, subfigure_increase.m

13.3 Review exercises and challenges

13.3.1 Review exercises

Review exercise R13-1

Obtain the expression for the Meissner currents in the dipole–dipole approximation, i.e. derive the sheet current $K_\infty^{\mathrm{dip}}(r)$ induced on a superconducting half-plane by a nearby small magnet (Figure 13.2).

Hint: use the expression for the magnetic field of the dipole and the condition of discontinuous tangential component at the surface.

Review exercise R13-2

Show that the generating function $g(t)$ for a dipole source $(0,0,m_0)$ at a distance a above a superconducting disk reads

$$g(t) = \frac{m_0}{\pi\sqrt{\pi}}\left[\frac{3at}{(a^2 + t^2)^2} + \frac{t^3}{a(a^2 + t^2)^2} - \frac{t}{a(a^2 + t^2)}\right] \qquad (13.27)$$

Based on this expression, use equation (13.13) and the script integ_K.m so as to obtain the sheet current induced by a tiny magnet above a superconducting disk with radius R_s, say $K_{R_S}^{\mathrm{dip}}(r)$. Compare the predictions of $K_{R_S}^{\mathrm{dip}}(r)$ and $K_\infty^{\mathrm{dip}}(r)$ for different values of a/R_s, f.i.: 0.2: 0.2: 1.

Hint: the symbolic integration of equation (13.12) for a dipole magnetic field may be straightforwardly done with the help of Mathematica.

Review exercise R13-3

Replot figure 13.4 using small magnets ($R_M/R_S = h_M/R_S = 0.1, 0.5, \ldots$) so as to make clear that the uniform field approximation clearly fails for $R_M/R_S < 1$ unless for the highest distances.

Hint: use the scripts related to figure 13.4.

Review exercise R13-4

Show that relation (13.8) can be inverted so as to obtain the current density in terms of the function $A(Rs, k)$, i.e. derive equation (13.9):

$$K(r) = \int_0^\infty dk \left[k A(R_S, k) J_1(kr) \right]$$

Hint: consider the completeness relation for Bessel functions.

Review exercise R13-5

Show that the generating function $g(t)$ for a cylindrical magnet (radius R_M, height h_M) uniformly magnetised along its axis, at a distance a from the surface of a coaxial superconducting disk is

$$g(t) = \frac{M_0}{\sqrt{\pi}} \left[\sqrt{t^2 - (R_M^2 + L_2^2) + G_2} - \sqrt{t^2 - (R_M^2 + L_1^2) + G_1} \right] \qquad (13.28)$$

with:

$$G_i \equiv \sqrt{\left[(R_M - t)^2 + L_i^2 \right] \left[(R_M + t)^2 + L_i^2 \right]}$$

and $L_1 = a$; $L_2 = a + h_M$

Hint: it will be useful to write the field created by the magnet as a superposition of elementary loops of magnetisation current in the form:

$$b(r) = \frac{M_0 R_M}{2} \int_{L_1}^{L_2} \int_0^\infty dk \left[k J_0(kr) J_1(kR_M) e^{-k|z|} \right] \qquad (13.29)$$

For the symbolic integration of equation (13.12) one may use Mathematica in combination with the tables in the book by Gradshteyn and Ryzhik [14].

Review exercise R13-6

Expression (13.13c) gives the self-energy of the superconducting disk under external field, say U_S. Together with a permanent magnet they constitute a system with a total energy that must also include both self and interaction terms, say $U = U_S + U_M + U_{SM}$. Why is then the force obtained by taking the derivative indicated in (13.13d)?

Hint: write the magnetic energy in terms of **J** and **A** and consider the boundary conditions of the problem.

Review exercise R13-7

Figures 13.7 and 13.8 display the magnetic force between a small magnet and a superconducting tape obtained by numerical differentiation of the interaction energy, i.e. straightforwardly apply equation (13.18). Verify that one may also obtain such force by direct summation over the force between long filaments (equation (13.24)). Check each plot against the alternative.

Hint: use the script that calculates the potential of the magnet $|\psi_{Sm}\rangle$ and apply numerical differentiation to obtain $\partial_z\langle\psi_{Sm}|\mathbf{K}\rangle$ for each position.

13.3.2 Challenges

Exercise C13-1

In 1947 V Arkadiev [5] demonstrated the stable levitation of a small ferro-nickel aluminium magnet above the surface of a concave lead disk in a Dewar vessel over liquid helium. In contrast to the configuration in figure 13.7, the concave shape of the superconductor enables a situation of stable levitation (see figure 13.12 below). With the help of the software accompanying figure 13.7 the reader is invited to reproduce the new situation and investigate the effect of changing the curvature of the bowl, the height of levitation etc. This test shows that, although possible, stable levitation with superconducting materials acting as diamagnetic bodies is not always straightforward.

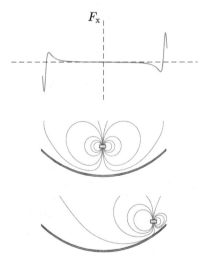

Figure 13.12. Lateral force on a small magnet floating above a bowl shaped superconductor (exercise C13-1).

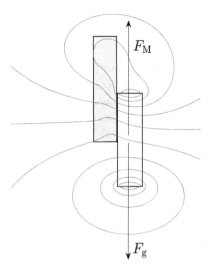

Figure 13.13. Demonstration of lateral suspension in a magnet/superconductor arrangement (exercise C13-2).

Exercise C13-2

One of the most striking demonstrations of levitation with type-II superconductors is the so-called 'lateral suspension'. As shown in figure 13.13, if one succeeds in creating a restoring vertical force that compensates gravity, and a zero or restoring horizontal force, one may have a magnet floating aside the superconductor. Starting with the software provided with figure 13.11, show that one may have $F_x < 0$ and $F_z < 0$ if the superconductor is cooled close to the magnet as shown in the plot.

Hint: consider the influence of the cooling position of the magnet and of the relative size of both elements.

References

[1] Moon F 2004 *Superconducting Levitation: Applications to Bearings and Magnetic Transportation* (Weinheim: Wiley)

[2] Navau C, del-Valle N and Sánchez A 2013 Macroscopic modeling of magnetization and levitation of hard type-II superconductors: The critical-state model *IEEE Trans. Appl. Supercond.* **23** 8201023

[3] Ma G T, Wang J S and Wang S Y 2010 3-D modeling of high-T_c superconductor for magnetic levitation/suspension application—Part I: introduction to the method *IEEE Trans. Appl. Supercond.* **20** 2219–27

[4] Ma G T, Wang J S and Wang S Y 2010 3-d modeling of high-T_c superconductor for magnetic levitation/suspension application—Part II: validation with experiment *IEEE Trans. Appl. Supercond.* **20** 2228–34

[5] Arkadiev V 1947 A floating magnet *Nature* **160** 330

[6] Jackson J D 1975 *Classical Electrodynamics* 2nd edn (New York: Wiley)

[7] Badía A 1997 Comment on magnetic levitation force and penetration depth in type-II superconductors *Phys. Rev.* B **55** 1875–6

[8] Badía A and Freyhardt H C 1998 Meissner state properties of a superconducting disk in a non-uniform magnetic field *J. Appl. Phys.* **83** 2681–8

[9] Sneddon I N 1972 *The Use of Integral Transforms* (New York: McGraw-Hill)

[10] Earnshaw S 1842 On the nature of the molecular forces which regulate the constitution of the luminiferous ether *Trans. Camb. Phil. Soc.* **7** 97–112

[11] Grover F W 1980 *Inductance Calculations* (Research Triangle Park, NC: Instrument Society of America)

[12] Kordyuk A A 1998 Magnetic levitation for hard superconductors *J. Appl. Phys.* **83** 610–2

[13] Badía-Majós A, Aliaga A, Letosa-Fleta J, Mora M A and Peña-Roche J 2015 Tradeoff modeling of superconducting levitation machines: theory and experiment *IEEE Trans. Appl. Supercond.* **25** 1–10

[14] Gradshteyn I S and Ryzhik I M 2007 *Table of Integrals, Series, and Products* (Amsterdam: Elsevier/Academic)

IOP Publishing

Macroscopic Superconducting Phenomena
An interactive guide
Antonio Badía-Majós

Chapter 14

Superconductors and magnets: cloaking devices

Along a line somehow different to the preceding chapters 12 and 13, here the emphasis will not be the interaction between magnets and superconductors, but their combination in the so-called heterostructures[1] so as to obtain unique properties enabled by their complementariness. In particular, we will show that integrating diamagnetic superconducting with paramagnetic layers, one may achieve static and low frequency cloaking of magnetic fields in some region of space. Such phenomena have been described by a number of authors in the last years, who provide theoretical, as well as experimental demonstrations of invisibility cloaks in the magnetostatic regime [1, 2, 3]. These works and others were inspired by a rather more ambitious program proposed in a pathbreaking work issued by J Pendry in 2006 [4]. Based on the property that the 'deformation' of the geometry of a region of space is equivalent to the modification of the underlying material properties (permittivity ϵ and permeability μ), he proposed that artificial metamaterials (tunable $\epsilon(\mathbf{r})$, $\mu(\mathbf{r})$) could be the path to modify the propagation of electromagnetic fields as desired in the region of interest. As explained by A Sánchez and co-workers, the practical construction of such a perfect electromagnetic cloak is virtually impossible because the constrictions on $\epsilon(\mathbf{r})$ and $\mu(\mathbf{r})$ in the region given, are very demanding. However, one may greatly simplify the construction of the cloak for magnetostatic fields by using a superconducting material combined with a homogeneous paramagnet. Owing to the high number of technological applications that rely on the existence of well-controlled magnetic fields, the possibility of manipulating their spatial distribution, blocking unwanted disturbances, etc is unquestionably appealing and has attracted many studies. Along this line, one may mention the

---

[1] In many instances, one may also find the term *metamaterials* for naming these structures in the sense that such artificial combinations of magnets and superconducting elements reach exclusive properties unexpected in their individual behaviour.

transportation of unperturbed magnetic field structures to long distances [5] or the drastic reduction of AC losses in power transmission lines [6].

Below, we will give the basic principles of cloaking devices based on super-conductor/paramagnet structures, as well as some practical tools for evaluation and simulation. Owing to the nature of the problem, the calculation methods will involve proficiency in two aspects.

- From the physical side, one will solve a complex system with simultaneous and interacting superconducting and magnetic responses[2].
- From the technical side, the high number of unknowns required to solve the problem in reasonable conditions will demand a refinement of the 'brute force' approximations that include each and all of the unknowns in the mathematical statement. Here, as done in chapter 10 for thin samples, we will introduce the concept of symmetry-aided reduction of the problem's size. Specifically, only one fourth of the number of unknowns will be used.

14.1 Pre-cloaking: magnetic shielding

A magnetic cloak is conceived as such a sheath that conceals an object while keeping the surrounding magnetic field unaffected. Also expected is that no distortion is produced within the cloak due to the external field. As emphasised in [1] and experimentally shown in [2, 3] such behaviour may be exactly achieved by a bilayer formed by a superconducting cylinder surrounded by magnetic material, as long as a uniform external field is concerned[3]. Below, we will concentrate on such bilayers. To start with let us analyse some connected physical phenomena (magnetic shielding) that will serve as a bridge to the formulation of a cloaking mechanism.

All along this chapter, we will consider long cylindrical structures extending along the z-axis subject to applied uniform magnetic fields in the perpendicular plane. We choose the representation $\mathbf{H}_a = H_0 \hat{\jmath}$ for such fields.

14.1.1 Shielding with paramagnets

Tightly related to cloaking is the concept of shielding of magnetic fields. As an example of this, we present figure 14.1 that illustrates a well known phenomenon. One may use high-permeability hollow cylinders so as to 'protect' enclosed objects from applied magnetic fields. Thus, the protecting shell 'absorbs' the magnetic field, leaving a flux free bore. Notice that shielding is nearly perfect for a relative permeability $\mu_r \equiv \mu/\mu_0$ above the value 100. Notice also, for further discussion, that the magnetic field lines within the cylinder for lower permeabilities remain straight for the circular cross section, while they get deformed for the tube of square section.

---

[2] Notice that in the previous chapters, the permanent magnet was considered unperturbed along the process (unaffected by the interaction with the superconductor).
[3] Cloaking of non-uniform magnetic fields (antimagnet behaviour [1]) may be achieved with a good degree of approximation by using multilayer structures.

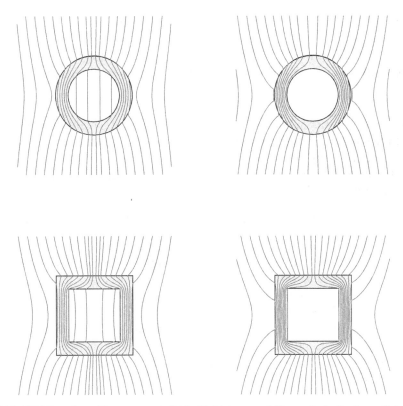

Figure 14.1. Shielding of a uniform magnetic field with high-permeability media in the form of circular and square tubes. Left panels: $\mu = 10\,\mu_0$, right panels: $\mu = 100\,\mu_0$. ✐. Source codes available at https://iopscience. iop.org/book/978-0-7503-2711-4.

Concerning the actual numerical implementation that allows to generate figure 14.1, the reader is addressed to section 14.2.1 where a general software is introduced that permits to model paramagnetic shielding as a particular case. Now, let us just introduce the notation and physical equations that give the basis for such calculations.

Along the lines introduced in sections 8.1.2 and 8.4, a *finite element* formulation has been considered. The magnetic domain will be characterised by a set of 'surface current density' elements $|\mathbf{K}_m\rangle$. Such elements will consist of strips of small width and infinite length along the cylinder axis that serve as a polygonal approximation of the magnet's boundaries. Recall that for a paramagnetic material *effective current densities* are described by surface layers at the interfaces, while volume current densities are zero[4]. For the case of our magnetic layer, a natural decomposition is $|\mathbf{K}_m\rangle \equiv |\mathbf{K}_{in}\rangle \oplus |\mathbf{K}_{out}\rangle$. Here, it is meant that for a magnetic sheath one has to

---

[4] The reader is invited to proof this property (exercise R14-1).

consider the superposition of surface current density elements both at the *inner* and *outer* boundaries of the tube[5].

Physically, in order to obtain the current distribution vector $|\mathbf{K}_{mag}\rangle$ one may recall the tangential continuity of the magnetic field vector at the interfaces:

$$B_t^+/\mu^+ = B_t^-/\mu^-. \tag{14.1}$$

Here $+$, $-$ are used as indicators of the position of two neighbouring points, one closely inside and one closely outside the respective boundary [7].

Equation (14.1) may be easily transformed into a linear system for the unknowns $|\mathbf{K}_m\rangle$ that may be solved numerically with ease. Thus, using Biot–Savart's law and superposition one obtains

$$(\mathbf{G}_{mag}^+ - \mu_r\,\mathbf{G}_{mag}^-)|\mathbf{K}_m\rangle = (\mu_r - 1)|\mathbf{H}_{0m}\rangle \tag{14.2}$$

Above, $\mathbf{G}$ represents a matrix that provides the tangential magnetic field values created by the current densities at the points of interest (now, the interfaces). On the other hand, $|\mathbf{H}_{0m}\rangle$ stands for the tangential component of applied magnetic field at points that define the surface of the magnet. In case of further contributions to the magnetic field, additional terms of the same nature should be included in the right hand side. The actual formulas that have been employed to obtain the matrix elements $\mathbf{G}_{mag}$ rely on the following expressions for the magnetic field created by a thin strip (of width L), stretching along the z-axis that carries a uniform surface current density, say K_s

$$B'_x(x', y') = -\frac{\mu_0 K_s}{2\pi}\left[\operatorname{atan}\frac{x' - L/2}{y'} - \operatorname{atan}\frac{x' + L/2}{y'}\right]$$

$$B'_y(x', y') = \frac{\mu_0 K_s}{4\pi}\ln\frac{(x' - L/2)^2 + y'^2}{(x' + L/2)^2 + y'^2} \tag{14.3}$$

Recall that, for notational simplicity, here relative coordinates $(x' \equiv x_i - x_j, y' \equiv y_i - y_j)$ are used. Eventually, one has $G_{ij} = \mathbf{B}(x_{ij}, y_{ij}) \cdot \boldsymbol{\tau}(x_i, y_i)$.[6]

In summary, the problem of paramagnetic field shields is solved by the numerical solution of the linear system (14.2) for some given geometry of the sheath (see exercise R14-3).

14.1.2 Shielding with superconductors

Another possibility for building a magnetic shield is illustrated in figure 14.2. Thus, if we use a superconducting (diamagnetic) shell, a field-free region may be created thanks to the flux expulsion phenomenon. Let us introduce the basic notation and equations for solving such problems. Two different situations will be considered as concerns the superconducting material. At first, and following the typical

---

[5] Recall the concept of *direct sum* introduced in chapter 5 (figure 5.5).
[6] Tangential component of the field created by the ith element at the jth position. Notice the local character of the unit vector that projects onto the tangential direction.

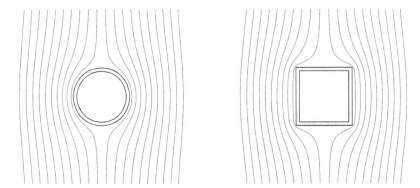

Figure 14.2. Shielding of magnetic fields by using superconducting layers. 🔗. Source codes available at https://iopscience.iop.org/book/978-0-7503-2711-4.

approximation in the literature of cloaking devices, we will treat the material as a perfect diamagnet. In such case, we will use the simplified variational principle (minimum energy) introduced in section 13.1.3 so as to simulate shielding configurations as those shown in figure 14.2. Here, the 'magnetic' potential $|\psi\rangle$ will be the related to a uniform applied field[7].

Then, equation (13.17) yields

$$|\mathbf{K}\rangle = -\mathbf{m}^{-1}\,|\psi_{\text{Applied}}\rangle \tag{14.4}$$

In the above equation $\mathbf{m}$ represents mutual inductance matrix between the superconducting elements. Being interested in the long tube geometry for the superconducting shell, these elements will be a set of infinite parallel wires[8], and again we will use equation (13.23) for evaluating their interaction terms.

Figure 14.2 shows the shielding property of a superconducting tube in the Meissner state and has been created on the basis of equation (14.4). The actual method for obtaining such plots is suggested in exercise R14-4. On the other hand, as the reader may foretell, equation (14.4) will have to be supplemented with additional terms when further contributions to the magnetic field are considered. This will be explained in the forthcoming section.

14.2 Cloaking bilayers: magnets and superconductors

From the above examples, it is clear that shielding capsules may be built either on the basis of paramagnetic or superconducting materials. Nevertheless, it is also clear that the existence of a 'protected region' may not be concealed, due to the deformation of the nearby field. Still, these examples also suggest an interpretation that will be useful for designing the magnetic cloak. One may notice that, while

---

[7] Recall (exercise R1-1), that for $\mathbf{H}_a = H_0\hat{\jmath}$, one has $|\psi_{\text{Applied}}\rangle = -(\mu_0 H_0)|\mathbf{x}\rangle$ with $|\mathbf{x}\rangle$ being the finite element vector giving the coordinates of the superconducting elements.

[8] Actually, one could refine it with a collection of thin tapes, giving a polygonal approximation of the perimeter.

'protecting' the inner region, the magnetic material and the superconductor produce reverse deformations (concave versus convex) of the magnetic field in the surrounding area. The question naturally arises: could a combination of both materials simultaneously shield the applied field within, as well as leave it undistorted in the nearby region?

14.2.1 Cloaking in the Meissner state

As explained and experimentally demonstrated in [2] the answer to the above question is affirmative, provided that a number of conditions are verified. Thus, for a cylindrical tube of magnetic material fulfilling the relation between the relative permeability and its inner and outer radii

$$\mu_r = \frac{R_2^2 + R_1^2}{R_2^2 - R_1^2}, \tag{14.5}$$

the (perfect)superconductor/paramagnet bilayer exactly works as a magnetic cloak for an applied uniform magnetic field of arbitrary magnitude. This property is illustrated in figure 14.3, that has been created for the case $R_2 = 1.5R_1 \Rightarrow \mu_r = 2.6$.

Figure 14.3 has been compiled by using equations (14.2) and (14.4) augmented with terms corresponding to the participation of each subsystem in the magnetic field acting on the other. Thus we solve the equations:

$$(\mathbf{G}_{\text{mag}}^{+} - \mu_r\, \mathbf{G}_{\text{mag}}^{-})|\mathbf{K}_{\text{m}}\rangle = (\mu_r - 1)|\mathbf{H}_{0m}\rangle + (\mu_r - 1)\mathbf{G}_{\text{sc}}|\mathbf{K}\rangle$$

$$|\mathbf{K}\rangle = -\,\mathbf{m}^{-1}\left(|\psi_{\text{Applied}}\rangle + |\psi_{\text{Magnet}}\rangle\right) \tag{14.6}$$

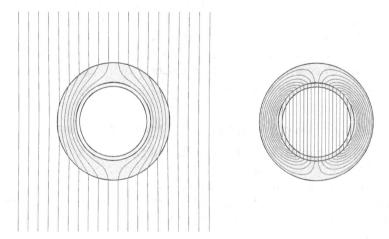

Figure 14.3. Magnetic cloak consisting of a superconducting layer in the Meissner state (green) and a surrounding paramagnetic shell (grey). To the left, we plot the full magnetic field, while, to the right we show the lines of field due to the bilayer. ⊘. Source codes available at https://iopscience.iop.org/book/978-0-7503-2711-4.

with $|\mathbf{K}_m\rangle$ and $|\mathbf{K}\rangle$ the unknowns[9], and $\mathbf{G}_{sc}$ the matrix that provides the tangential magnetic field created by the superconductor at the boundaries of the magnet. These elements are evaluated on the basis of the infinite wire approximation, i.e. from

$$\mathbf{B}_{sc}(x', y') = \frac{\mu_0 I_{sc}}{2\pi} \frac{(-y', x')}{x'^2 + y'^2} \tag{14.7}$$

with I_{sc} being the superconducting current along the elementary wire centred at the source point (x_j, y_j), and again (x', y') standing for relative *field-source* coordinates. As before, we use the relation $G_{ij} = \mathbf{B}(x_{ij}, y_{ij}) \cdot \boldsymbol{\tau}(x_i, y_i)$.

From the technical point of view, the system (14.6) may be solved in a number of ways. In the accompanying software, we use a straightforward process relying on MATLAB scripts that is solely based on elementary matrix algebra (inversion and multiplication). The reader is addressed to exercise R14-5 for some guide on the steps to be taken. Below, we list the scripts that perform such operations, most of which are common to those provided for figure 14.4 and described at the end of this chapter.

- `cloak_cyl_meiss.m` (main script for figure 14.3)
- `define_param_meiss.m`, `generate_mesh_meiss.m`
- `B_xy_het_p.m`, `B_xy_sc.m`, `B_xy_m.m`, `B_app_project.m`
- `matrixM_II_par.m`, `mutual.m`
- `vector_field_plot_A.m`, `A_xy_g_sc.m`, `A_xy_g_m.m`

14.2.2 Cloaking in the critical state

In this section, we deal with an interesting property: cloaking conditions may be also met for bilayers consisting of a paramagnet and a superconductor in the critical state. The relevance of this result stems from the fact that the range of operation noticeably enlarges (to the range of tesla in the magnetic flux density) as compared to the above considered Meissner state cloaking. Nevertheless a relevant short-coming will appear. In contrast to such case, now cloaking will only occur for a specific value of the magnetic field applied to the heterostructure.

Some analytical results

As shown by Yampolskii and Genenko [8] one may issue a number of analytical predictions related to the cloaking of magnetic fields with superconductors in the critical state. In particular, these authors give formulas for the applied magnetic field at which full penetration (H_{fp}) and cloaking (H_{cloak}) occur in a cylindrical bilayer. Their basic assumption was that $H_{fp} < H_{cloak}$, i.e. cloaking occurs in the *fully penetrated* state. As expected, these characteristic fields depend on the geometry of

---

[9] Recall that $|\psi_{Magnet}\rangle$ may be written in terms of $|\mathbf{K}_m\rangle$, *viz.*, as given by equation (14.12) and may be computed through the function `A_xy_g_m.m`.

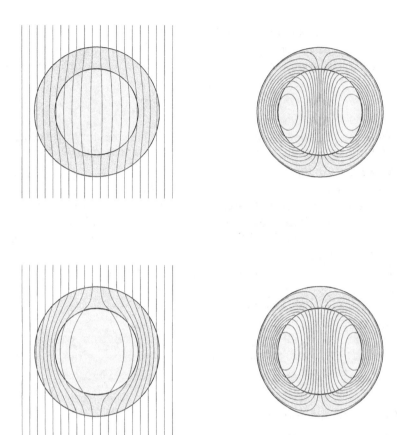

Figure 14.4. Cloaking of magnetic fields by using small permeability material (outer shell) in combination with a superconductor in the critical state (inner cylinder). Upper panels: $\mu_r = 1.2$, lower panels: $\mu_r = 2$. To the left, we show the full magnetic field structure, to the right the contribution of the heterostructure (see text). 🔗. Source codes available at https://iopscience.iop.org/book/978-0-7503-2711-4.

the bilayer (inner and outer radii of the paramagnet R_1, R_2), as well as on the material properties μ_r and J_c. Specifically

$$H_{fp} = \frac{2J_c R_1}{\pi}\left[1 + \frac{(\mu_r - 1)(2\mu_r - 1)(1 - R_1^2/R_2^2)}{6\mu_r}\right] \quad (14.8)$$

$$H_{cloak} = \frac{8J_c R_1}{3\pi}\frac{\mu_r}{\mu_r^2 - 1}\frac{R_1^2}{R_2^2 - R_1^2} \quad (14.9)$$

The above expressions correspond to the case of a paramagnetic sheath tightly covering a superconducting cylinder of radius R_1. They can be used as a guide for solving the problem concerning the properties of a bilayer that cloaks a given magnetic field.

Numerical statement

The possibility of encountering cloaking conditions with the superconductor in the *partial penetration regime*[10] was investigated numerically in [9]. As an important result of that work, it was shown that, for a given geometry, cloaking may also occur in a range of parameter values in between the bounds established by equations (14.5) and (14.8)–(14.9). Thus, physically, for a given ratio R_2/R_1 one has a maximum μ_{max} (given by (14.5)). On the other hand, a minimum value μ_{min} (determined by the condition $H_{fp} < H_{cloak}$) is required to warrant cloaking in the full penetration regime. What occurs in the intermediate range $\mu_{min} < \mu < \mu_{max}$ is that one may find a cloaking field with partial penetration in the superconductor.

In order to study the related phenomena, we provide the software resources that may be used to generate figure 14.4. There, we illustrate the appearance of cloaking in the full and partial penetration regimes. The basic structure of the underlying numerical procedure is as follows. Let us assume that the heterostructure is subject to some process given by the evolution of the applied magnetic field: $H_0(t_n) \rightarrow H_0(t_n + \Delta t)$. By using the notation and definitions introduced in chapter 8 (equation (8.26)), and according to [9], the coupled evolution of the heterostructure may be obtained through to the following double step incremental process

$$
\begin{array}{l}
|J^\vee\rangle \quad \text{and} \quad |K_m^\vee\rangle \text{ are given at time } t_n, \text{ then:} \\
\hline
\begin{array}{l} |J^\vee\rangle \\ H_0(t_n + \Delta t) \end{array} \xrightarrow{\text{paramagnet's law}} |K_m\rangle \\
\hline
\begin{array}{l} |K_m\rangle \\ H_0(t_n + \Delta t) \end{array} \xrightarrow{\text{superconducting law}} |J\rangle \\
\hline
\text{and so on}
\end{array}
\tag{14.10}
$$

Above, $|\mathbf{J}^\vee\rangle$ is used for the current density finite element vector at the points of the superconductor, at the time step t_n (analogously for $|\mathbf{K}_m^\vee\rangle$ in the magnet), while $|\mathbf{J}\rangle$ and $|\mathbf{K}_m\rangle$ are their updated values for $t = t_n + \Delta t$.

For each step in the process, the paramagnet's and superconducting law are solved according to

$$
(\mathbf{G}_{mag}^+ - \mu_r\,\mathbf{G}_{mag}^-)|\mathbf{K}_m\rangle = (\mu_r - 1)|\mathbf{H}_{0m}(t_n + \Delta t)\rangle + (\mu_r - 1)\mathbf{G}_{sc}|\mathbf{J}^\vee\rangle
\tag{14.11}
$$

with unknown $|\mathbf{K}_m\rangle$, and then, we solve

$$
\texttt{minimise} \quad \left[\frac{1}{2}\langle \mathbf{J}|\mathbf{m}|\mathbf{J}\rangle - \langle \mathbf{J}^\vee|\mathbf{m}|\mathbf{J}\rangle + \langle \Delta\psi_{Applied}|\mathbf{J}\rangle + \langle \mathbf{K}_m - \mathbf{K}_m^\vee|\mathbf{P}|\mathbf{J}\rangle\right]
\tag{14.12}
$$

with unknown $|\mathbf{J}\rangle$.

We notice that in equation (14.11) the interaction of the paramagnet and the superconductor is stated in terms of a mutual induction matrix $\mathbf{P}$ that couples the current density elements within the magnet and the superconductor. Thus, the

---

[10] See section 3.3 for the extended description of field penetration in hard type-II superconductors.

coefficient that couples the ith magnetic surface current element (long tape of width L) to the jth superconducting element reads

$$P_{ij} = \frac{\mu_0}{4\pi} \left\{ (x_{ij} - L)\ln[(x_{ij} - L)^2 + y_{ij}^2] + (x_{ij} + L)\ln[(x_{ij} + L)^2 + y_{ij}^2] \right.$$
$$\left. - 2y_{ij}\left[\mathrm{atan}\frac{L - x_{ij}}{y_{ij}} + \mathrm{atan}\frac{L + x_{ij}}{y_{ij}} \right] \right\} \tag{14.13}$$

Detection of the cloaking condition

By means of equations (14.10)–(14.12) one may investigate the response of a given heterostructure to the applied magnetic field. One may, for instance, focus on the influence of μ_r in the appearance of the cloaking condition. For this purpose, one may obtain the response $|\mathbf{K}_m\rangle(t)$ and $|\mathbf{J}\rangle(t)$, plot the lines of the full magnetic field and check. Alternatively, one may plot the field created by the heterostructure itself (right panels of figure 14.4). It is obvious that when cloaking occurs, these lines do not leak out. Nevertheless, and considering that this may be a rather tedious task because considerable precision is needed, below we will introduce an accurate and fast characterisation method previous to the plot of the lines of field. As argued in [8], the invisibility of the bilayer also means that the total magnetic moment goes to zero. This condition is straightforwardly checked by applying equation (8.28) to the full set of currents, i.e. $|\mathbf{K}_m\rangle \oplus |\mathbf{J}\rangle$. Figure 14.5 illustrates the determination of H_{cloak} by this method. Additionally, we plot the magnetic moment of the superconducting part, whose saturation gives the full penetration field predicted by equation (14.8).

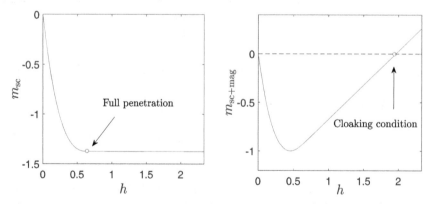

Figure 14.5. Behaviour of the magnetic moment of the heterostructure as a function of the applied magnetic field. Left: the contribution of the superconductor. Right: the complete magnetic moment becomes zero at the cloaking condition. We have used $R_2 = 1.5R_1$ and $\mu_r = 1.2$. Units for h are J_cR_1 and for the magnetic moment are $J_cR_1^2$. 🖉. Source codes available at https://iopscience.iop.org/book/978-0-7503-2711-4.

Notice that, above we just imply that the dipole term in the multipole expansion goes to zero. In fact, the absence of higher multipoles is ensured in our cylindrical shell under uniform field. This property relates to the appearance of a uniform field within the paramagnetic structure as said in section 14.1.1 and may be also exploited for the spheroidal geometry.

Numerical resources

Next, we will describe the scripts that have been used to produce the figures of this chapter, and may be the basis for further studies. Indeed, we provide the means for creating figure 14.4, that deals with the most general situation. As sketched in figure 14.6, the cross section of the cylindrical heterostructures is meshed by (i) a grid of points that cover the superconducting region (current lines j_i), and (ii) a set of parallel tapes along the axis that provide a polygonal approximation of the boundaries of the magnet (surface currents $k_{m,\alpha}$). This forms the basis for the application of equations (14.10)–(14.12). Specifically, we do it with the scripts listed below:

- cloak_cyl.m (main script for figure 14.4)
- define_param.m, generate_mesh.m, replicate.m
- B_xy_het_p.m, B_xy_sc.m, B_xy_m.m, B_app_project.m
- update_magnet.m
- matrixM_II_par.m, matrixM_IJ_par.m, mutual.m
- magnetic_moment.m
- vector_field_plot_A.m, A_xy_g_sc.m, A_xy_g_m.m

The main program creates the specific problem through some auxiliary scripts that build the mesh of points defining the geometry of the heterostructure and its physical parameters (μ_r, J_c).

The application of equation (14.11), i.e. the physical model for the paramagnetic shell is performed through the script update_magnet.m, that uses the tangential components of the different contributions to the magnetic field, supplied by the

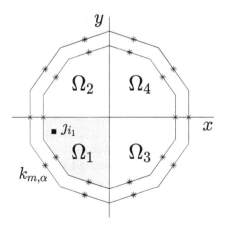

Figure 14.6. Sketch of the finite element description of the heterostructure: long cylinder along z-axis. The cross section is divided into four sectors.

functions `B_xy_full_p.m`, `B_xy_sc`, `B_xy_m.m`, `B_app_project.m`. It is of mention that the magnetic field is evaluated at the set of points that add up to define the full boundary of the magnetic layer. This includes N_i points for the inner surface, and N_o for the outer surface. Thus, in particular, $\mathbf{G}_{mag}$ must be a $(N_i + N_o) \times (N_i + N_o)$ matrix, that multiplied by the current density vector $|\mathbf{K}_m\rangle$, i.e. $|\mathbf{K}_{in}\rangle \oplus |\mathbf{K}_{out}\rangle$ gives the tangential component of the field at the points of the boundary. By applying the relation (14.3) to the conveniently sorted set of combinations (x_{ij}, y_{ij}) one gets the matrix $\mathbf{G}_{mag}$, that takes the form[11]

$$\mathbf{G}_{mag} \rightarrow \left[\begin{array}{c|c} N_i \times N_i & N_i \times N_o \\ \hline N_o \times N_i & N_o \times N_o \end{array} \right] \tag{14.14}$$

*Concerning the application of equation (*14.12*)*, a key element of the simulation is that, for the cases under study, one may reduce the number of unknowns to one fourth of the grid. As one may check through the figures along the chapter, three quarters of the problem may be obtained by 'replicating' the structure of the remaining one. This property has been used to accelerate the solution of equation (14.12) for a relatively high number of unknowns (10 000 elements may be dealt with in a personal computer), and thus obtain rather accurate representations of the lines of field. In particular, taking advantage of the relation[12]

$$J(x, y) = J(x, -y) = -J(-x, y) = -J(-x, -y) \tag{14.15}$$

one may perform the minimisation step just solving for the unknowns in one quadrant (as schematically shown in figure 14.6). Nonetheless, as the reader may guess, one should not skip the fact that the elementary current wires of the full section interact all with each other through the coupling matrix elements m_{ij}.[13] Actually, this physical requirement is fulfilled by considering equation (14.15). Then (see exercise R14-7) one may show that

$$\langle \mathbf{J}|\mathbf{m}|\mathbf{J}\rangle = 4(\langle \mathbf{J}_1|\mathbf{m}^{11}|\mathbf{J}_1\rangle + \langle \mathbf{J}_1|\mathbf{m}^{12}|\mathbf{J}_1\rangle \\ - \langle \mathbf{J}_1|\mathbf{m}^{13}|\mathbf{J}_1\rangle - \langle \mathbf{J}_1|\mathbf{m}^{14}|\mathbf{J}_1\rangle) \tag{14.16}$$

Here, $|\mathbf{J}_1\rangle$ stands for the set of current elements in the region Ω_1, and the matrices $\mathbf{m}^{11}$, $\mathbf{m}^{12}$, $\mathbf{m}^{13}$, $\mathbf{m}^{14}$ couple the points of the region Ω_1 with Ω_1, Ω_2, Ω_3, Ω_4 respectively. The actual implementation of equation (14.16) is performed through the functions `matrixM_II_par.m`, `matrixM_IJ_par.m`, `mutual.m`.

Eventually, having obtained the full current density vector $|\mathbf{K}_m\rangle \oplus |\mathbf{J}\rangle$, one may evaluate a number of physical quantities. In particular, the different contributions to the dipole magnetic moment may be evaluated through `magnetic_moment.m`.

---

[11] A similar argument can be applied to the matrix $\mathbf{G}_{sc}$, but now relying on equation (14.7) and the superconducting current vector $|\mathbf{J}\rangle$.

[12] If one sets the origin of coordinates at the centre of the heterostructure.

[13] m_{ij} couples the elements j_i and j_j with locations (x_i, y_i) and (x_j, y_j).

Also, once again, we exploit the property that the lines of $\mathbf{B}$ may be obtained as isolines of the vector potential, i.e. $A_z(x, y) = $ constant. This is performed through the codes vector_field_plot_A.m, A_xy_g_sc.m, A_xy_g_m.m, that evaluate the full vector potential in some specific region that contains the heterostructure. For the readers' sake, the additional file figure_14_4.m is also included in section so as to provide a guide for making the plots.

14.3 Review exercises and challenges

14.3.1 Review exercises

Review exercise R14-1

Show that a linear magnetic material subjected to some applied magnetic field related to external sources does not develop effective volume current densities as a response to the excitation.

Hint: evaluate $\nabla \times \mathbf{M}$ with $\mathbf{M}$ its magnetisation.

Review exercise R14-2

The material law for paramagnetic layers has been introduced by recalling the tangential continuity of the magnetic field strength $\mathbf{H}$, that leads to the condition (14.1) for the magnetic flux density $\mathbf{B}$. Nothing has been mentioned concerning the continuity (or not) of the normal component of $\mathbf{B}$, i.e. $B_n^+ = B_n^-$, which should certainly be guaranteed in any model. Show that, in fact, this condition is automatically satisfied owing to the nature of our method.

Hint: material laws are introduced in terms of either real or effective current densities, which are the sources of the related magnetic flux density.

Review exercise R14-3

Generate the plots in figure 14.1

Hint: figure 14.1 relies on the application of equation (14.2), which is nothing but a particular case of the system (14.6) for $|\mathbf{K}\rangle = 0$.

Review exercise R14-4

Generate the plots in figure 14.2

Hint: figure 14.2 relies on the application of equation (14.4), which is nothing but a particular case of the system (14.6) for $|\mathbf{K}_m\rangle = 0$.

Review exercise R14-5

Show that the problem of a cloaking bilayer with the superconductor in the Meissner state may be solved straightforwardly through matrix multiplication.

Hint: recall that in equation (14.6) $|\psi_{\text{Magnet}}\rangle$ may be expressed in terms of $|\mathbf{K}_m\rangle$ through a linear relation, say $|\psi_{\text{Magnet}}\rangle = \mathbf{A}_m|\mathbf{K}_m\rangle$, and $|\mathbf{K}_m\rangle$ written in terms of $|\mathbf{H}_{0m}\rangle$ and $|\mathbf{K}\rangle$. Then, back substitution and some matrix algebra lead to $|\mathbf{K}\rangle = \mathbf{M}_{\text{msc}}|\psi_{\text{Applied}}\rangle$, with $\mathbf{M}_{\text{msc}}$ a matrix that incorporates the action of all the underlying sources, both magnetic and superconducting. The actual matrix operations may be checked by the reader following the structure of the main program `cloak_cyl_meiss.m`. Recall also that a valid expression for the vector potential of a uniform magnetic field along the y-axis is given by $|\psi_{\text{Applied}}\rangle = -\mu_0 H_0|\mathbf{x}\rangle$.

Review exercise R14-6

Related to figure 14.5: check the robustness of the 'cloaking condition', i.e. shift the applied field by some percentage around the value H_{cloak} and study the breakdown of the condition $m_{\text{sc+mag}} = 0$, together with the actual deviation of the magnetic field lines around the heterostructure.

Hint: use the scripts related to figure 14.4.

Review exercise R14-7

Justify relation (14.16).

Hint: in principle, either with symmetry or not, inductive coupling must be considered between the full set of current elements in the superconducting region. Formally (recall figure 14.6), one must calculate $\langle \mathbf{J}|\mathbf{m}|\mathbf{J}\rangle$ being $|\mathbf{J}\rangle = |\mathbf{J}_1\rangle \oplus |\mathbf{J}_2\rangle \oplus |\mathbf{J}_3\rangle \oplus |\mathbf{J}_4\rangle$ with $|\mathbf{J}_1\rangle \in \Omega_1$ and so on. However, due to symmetry, the sets of values $|\mathbf{J}_1\rangle, |\mathbf{J}_2\rangle, |\mathbf{J}_3\rangle, |\mathbf{J}_4\rangle$ are coincident up to a sign. Then, one may reorganise the full summation as indicated, just by picking the corresponding coefficients of the full matrix $\mathbf{m}$.

14.3.2 Challenges

Exercise C14-1

By changing the mutual inductance coefficients to the case of circular loops, recreate figure 14.3 for the case of a spherical bilayer cloak. Verify that equation (14.5) is replaced by equation (S21) in [2], i.e.

$$\mu_r = \left(2R_2^3 + R_1^3\right)/2\left(R_2^3 - R_1^3\right)$$

Exercise C14-2

Investigate the cloaking condition for a hollow cylindrical superconductor in the critical state. Compare your results to those in [8].

References

[1] Sánchez A, Navau C, Prat-Camps J and Chen D X 2011 Antimagnets: controlling magnetic fields with superconductor-metamaterial hybrids *New J. Phys.* **13** 093034

[2] Gömöry F, Solovyov M, Šouc J, Navau C, Prat-Camps J and Sánchez A 2012 Experimental realization of a magnetic cloak *Science* **335** 14664

[3] Šouc J, Solovyov M, Gömöry F, Prat-Camps J, Navau C and Sánchez A 2013 A quasistatic magnetic cloak *New J. Phys.* **15** 053019

[4] Pendry J B, Schurig D and Smith D R 2006 Controlling electromagnetic fields *Science* **312** 1780–2

[5] Navau C, Prat-Camps J, Romero-Isart O, Cirac J I and Sánchez A 2014 Long-distance transfer and routing of static magnetic fields *Phys. Rev. Lett.* **112** 253901

[6] Genenko Y A, Rauh H and Kurdi S 2015 Finite-element simulations of hysteretic alternating current losses in a magnetically coated superconducting tubular wire subject to an oscillating transverse magnetic field *J. Appl. Phys.* **117** 243909

[7] Jackson J D 1975 *Classical Electrodynamics* 2nd edn (New York: Wiley)

[8] Yampolskii S V and Genenko Y A 2014 Magnetic cloaking by a paramagnet/superconductor cylindrical tube in the critical state *Appl. Phys. Lett.* **104** 143504

[9] Peña-Roche J, Genenko Y A and Badía-Majós A 2016 Magnetic invisibility of the magnetically coated type-II superconductor in partially penetrated state *Appl. Phys. Lett.* **109** 092601

IOP Publishing

Macroscopic Superconducting Phenomena
An interactive guide
Antonio Badía-Majós

Chapter 15

Intermediate Josephson junctions

The main focus of this book are the *macroscopic* manifestations of superconductivity, and thus, we have provided a compilation of utilities connected to the realm of Classical Electromagnetics. Nonetheless, as was emphasised in chapter 3, the quantum nature of the phenomenon reveals distinctly in the electrical properties of Josephson junctions. Not only for this, but also for the wide range of alluring applications that are already at hand or under development, it seems convenient to dedicate this chapter to such effect.

In chapter 3 (section 3.4) we provided a basic introduction to the physics of Josephson junctions. Idealised configurations were used so as to reveal the main phenomena through relatively simple analytical calculations. Here we put a focus on more realistic models that allow to go deeper into the problem assisted by numerical tools. This is meant by 'intermediate Josephson junctions'.

A good number of excellent books have devoted full or partial attention to the subject and are strongly recommended [1, 2, 3] for a wider perspective. Here, we will concentrate on two milestones, i.e. the response to applied magnetic fields and the properties of the resistive transition. A collection of numerical resources which may introduce the reader to these fascinating hallmarks are provided.

15.1 Critical currents in planar Josephson junctions

Josephson junctions in planar geometry have attracted much interest, especially since the advent of nanofabrication techniques. As sketched in figure 15.1, by planar junction we mean the edge contact between two non-overlapping superconducting films. Although the underlying physics does not differ much from the elements used in chapter 3 to describe bulk junctions, the quasi-2D nature of the problem requires further consideration. A realistic and meaningful physical model should incorporate the effect of finite geometry. In particular, concerning the $I_c(B)$ behaviour, one needs to upgrade the 1D model presented in section 3.4. As we will see below, stray fields may noticeably determine the behaviour of the *gauge invariant phase difference*

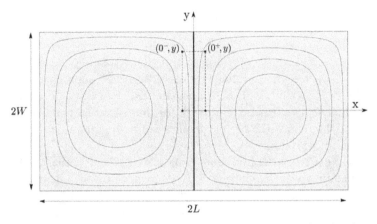

Figure 15.1. Sketch of the planar superconducting sample with an edge-type Josephson junction at the middle. Also shown are the screening current density streamlines induced by a uniform magnetic field applied perpendicular to the plane of the sample. ✐. Source codes available at https://iopscience.iop.org/book/978-0-7503-2711-4.

along the junction. Recalling Josephson's supercurrent relation, i.e. $J = J_c \sin(\Delta\varphi)$ this will have an apparent influence on the pattern $I_c(B)$.

15.1.1 Constitutive equations

A number of works have considered the problem sketched in figure 15.1. The edge-type junction acts as a barrier between two superconducting plates. Here, the analysis will proceed along the lines established by J R Clem [4]. To start with, the physical model has to incorporate the electromagnetic quantities, as well as the superconducting order parameter. The second Ginzburg–Landau equation will provide these elements.

Let us start by recalling such equation:

$$\mathbf{J} = -f^2 \frac{\Phi_0}{2\pi\mu_0\lambda_L^2}\left(\nabla\theta + \frac{2\pi}{\Phi_0}\mathbf{A}\right) \tag{15.1}$$

As shown by Clem, it may be cast in the more convenient form for the 'quasi-2D' geometry[1]:

$$\mathbf{K} = -f^2 \frac{\Phi_0}{2\pi\mu_0\Lambda}\left(\nabla\theta + \frac{2\pi}{\Phi_0}\mathbf{A}\right) \tag{15.2}$$

with $\mathbf{K}$ being the 'sheet current' across the sample's thickness d

---

[1] 'Quasi-2D' means that one can solve the problem in the sample's XY plane though a magnetic field in the z direction be applied.

$$\mathbf{K}(x, y) = \int_{-d/2}^{d/2} \mathbf{J}(x, y, z) dz, \tag{15.3}$$

with θ being the superconductor's phase function, $\mathbf{A}$ being the magnetic vector potential and f being the modulus of the 'relative' superconducting's order parameter $|\Psi|/|\Psi_\infty|$, which will be approximated by unity in what follows[2]. $\Lambda \equiv \lambda_L^2/d$ is the *effective* penetration depth for thin films. Notice that one assumes that the extreme thinness of the sample justifies averaging quantities across z without loss of essential information.

We call the readers' attention that equation (15.2), together with symmetry arguments (including both geometry and gauge invariance), will help to solve for the behaviour of the superconductor with the tunnel junction. To start with, as emphasised in section 3.4, we recall the relevant expression of the gauge invariant phase difference:

$$\Delta\varphi = \Delta\theta - \frac{2\pi}{\Phi_0} \int \mathbf{A} \cdot d\ell \tag{15.4}$$

that applies between any two points of the superconductor. Remarkably, it also acquires sense across the junction as long as tunnelling is involved. Let us consider the circuit in figure 15.1 that embraces the junction[3]. One has

$$\varphi(0^+, y) - \varphi(0^+, 0) = \theta(0^+, y) - \theta(0^+, 0) - \frac{2\pi}{\Phi_0} \int_{(0^+,0)}^{(0^+,y)} A_y \, dy$$

$$\varphi(0^-, y) - \varphi(0^-, 0) = \theta(0^-, y) - \theta(0^-, 0) - \frac{2\pi}{\Phi_0} \int_{(0^-,0)}^{(0^-,y)} A_y \, dy \tag{15.5}$$

Subtracting, and using $\Delta\varphi(y) \equiv \varphi(0^+, y) - \varphi(0^-, y)$ we get

$$\Delta\varphi(y) \approx \Delta\varphi(0) + [\theta(0^+, y) - \theta(0^+, 0)] - [\theta(0^-, y) - \theta(0^-, 0)] \tag{15.6}$$

Notice that here, recalling an extremely narrow circuit, we have used continuity of the vector potential. In contrast, the phase function θ remains, i.e. $\theta(0^+, y) \not\approx \theta(0^-, y)$ because one is in fact dealing with two different media.

Finally, the bracketed quantities may be evaluated with the help of equation (15.2) within each superconductor (with the assumption $f \approx 1$)

$$\theta(0^+, y) - \theta(0^+, 0) = -\frac{2\pi\mu_0\Lambda}{\Phi_0} \int_{(0^+,0)}^{(0^+,y)} K_y \, dy - \mu_0\Lambda \int_{(0^+,0)}^{(0^+,y)} A_y \, dy$$

$$\theta(0^-, y) - \theta(0^-, 0) = -\frac{2\pi\mu_0\Lambda}{\Phi_0} \int_{(0^-,0)}^{(0^-,y)} K_y \, dy - \mu_0\Lambda \int_{(0^-,0)}^{(0^-,y)} A_y \, dy \tag{15.7}$$

---

[2] In the sense that the experimental conditions are assumed to be such that a negligible reduction of f is produced.
[3] The notation $x = 0^\pm$ refers to the vertical lines, infinitesimally close to the Y axis ($x = 0$), either to the right or to the left.

Then

$$\Delta\varphi(y) \approx \Delta\varphi(0) + \frac{4\pi\mu_0\Lambda}{\Phi_0} \int_{(0^+,0)}^{(0^+,y)} K_y(0^+, y)\, dy \qquad (15.8)$$

where we have used the symmetry condition

$$K_y(0^-, y) = -K_y(0^+, y) \qquad (15.9)$$

and again neglected the line integral of A_y, corresponding to a negligibly small magnetic flux threading through the rectangle[4].

Thus, upon knowing $K_y(0^+, y)$ one may use the above equation and eventually solve for I_c by integrating Josephson's sinusoidal relation along the junction. A typical route to work this out entails going back to equation (15.2). Notice that, if one takes the **curl** of both sides, the phase function 'disappears' and, as long as $f^2 \approx 1$, one gets the second London equation

$$\mu_0\Lambda\nabla \times \mathbf{K} + \mathbf{B} = 0 \qquad (15.10)$$

In the forthcoming section, we put forward the solution of equation (15.10) that allows to trace back and obtain the pattern of the current density displayed in figure 15.1

15.1.2 The integrated current density

The above equation may be transformed into scalar form by recalling the condition of charge conservation, i.e. $\nabla \cdot \mathbf{K} = 0$. Mathematically, this allows us to introduce the so-called *stream function*[5] $\sigma(x, y)$ such that

$$\mathbf{K} = \frac{\Phi_0}{2\pi\mu_0\Lambda}\nabla \times (\sigma\hat{\mathbf{k}}) \qquad (15.11)$$

Then, London's equation takes the form

$$\frac{\partial^2\sigma}{\partial x^2} + \frac{\partial^2\sigma}{\partial y^2} = \frac{2\pi}{\Phi_0}B_z \qquad (15.12)$$

In principle, this is still a complex problem, but it can be further simplified under some rather plausible approximations that allow to extract the basic physics of the problem, *viz.* the influence of the screening currents in the response of the junction to the magnetic field. The equation will be solved under the conditions:

- The tunnelling currents are weak and one may assume that σ is mostly a screening contribution with tangential flow along the boundaries.
- The self-field generated by the screening currents is weak as compared to the applied field, as corresponds to the sample being very thin (i.e. $\Lambda \gg L, W$) [4]

---

[4] Recall the relation $\oint \mathbf{A} \cdot d\ell = \Phi$.
[5] Recall the *stream function formalism* introduced in chapter 10.

which allow to assume that: (i) σ is constant along the boundaries (recall that $K_x = \partial_y\sigma$, $K_y = -\partial_x\sigma$), and (ii) B_z in the above equation may be identified with the applied magnetic field $B_z \approx B_{applied} \equiv B$.

Thus, equation (15.12) becomes Poisson's equation. As shown in [4], its solution may be expressed as follows

$$\sigma(x, y) = \frac{2\pi}{\Phi_0} B \left[y^2 + \frac{8}{W} \sum_{n=0}^{\infty} \frac{(-1)^n \cosh[k_n(x - L/4)]\cos(k_n y)}{k_n^3 \cosh(k_n L/4)} \right] \qquad (15.13)$$

with $k_n \equiv \pi(2n + 1)/W$.

The above expression was used to create figure 15.1. A fast convergence is ensured as one may check through the supplied MATLAB code. Recall that the plot of the current density streamlines is based on the property of the stream function: $\mathbf{K} \cdot \nabla\sigma = 0$. Apparently, this implies that the contour levels of σ (perpendicular to its gradient) give us the streamlines of the current density.

15.1.3 The field dependence $I_c(B)$

Equations (15.8) and (15.13) together with the definition of σ may now be used to obtain the gauge invariant phase difference between the banks of the junction, and eventually the field dependence of the critical current. Thus, one gets

$$\Delta\varphi(y) = \Delta\varphi_0 + \frac{16\pi B}{\Phi_0 W} \sum_{n=0}^{\infty} \frac{(-1)^n \tanh(k_n L/4)\sin(k_n y)}{k_n^3} \qquad (15.14)$$

$$\equiv \Delta\varphi_0 + \Delta\varphi_B(y)$$

Let us analyse this expression and the subsequent critical current profile $I_c(B)$. Recall that the maximum superconducting tunnelling current across the junction is given by

$$\max\{I(B)\} = K_c(0) \max \left\{ \int_{-W/2}^{W/2} \sin[\Delta\varphi_B(y)]dy \right\} \qquad (15.15)$$

with $K_c(0)$ being the zero-field critical sheet current density of the junction. This follows from Josephson's relation for the current density (section 3.4).

Now, the above maximum value may be found as follows[6]

$$I_c(B) = I_c(0)\frac{2}{W} \left| \int_0^{W/2} \cos[\Delta\varphi_B(y)]dy \right| \qquad (15.16)$$

This expression has been implemented in our scripts, and together with equation (15.14) has been used to obtain figure 15.2. The function $I_c(B)/I_c(0)$ is shown for the 'short' strip case $W = 2L$, as well as $\Delta\varphi_B(y)$ for several values of the aspect ratio W/L. Concerning the magnetic field, as pointed out in the inset, for each case, we have chosen a characteristic value $B_1(W/L)$ defined as follows. It is the lowest value of the

---

[6] The proof of this property will be proposed as a guided exercise (R15-2) at the end of this chapter.

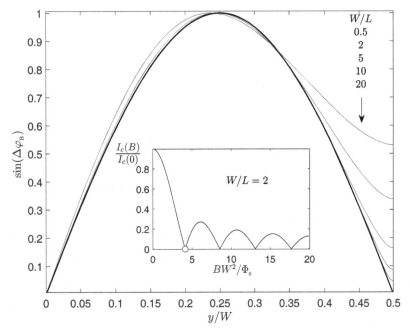

Figure 15.2. Sinusoidal dependence of the gauge invariant phase variation $\Delta\varphi_B$ across the planar junction in figure 15.1 as a function of the vertical coordinate y for various values of the aspect ratio W/L. The applied magnetic field for all cases corresponds to the first minimum in the $I_c(B)$ dependence. The inset illustrates the determination of such point for $W/L = 2$. For reference, the thick line represents the limiting case of sinusoidal dependence $(\Delta\varphi_B(y) = \sin(2\pi y/W))$ for $W/L \gg 1$. $\mathscr{O}$. Source codes available at https://iopscience.iop.org/book/978-0-7503-2711-4.

field for which I_c goes to zero, and as indicated, depends on W/L. In fact, as estimated from the 1D approximation (chapter 3), the formation of a single Josephson vortex $(\Phi_J = \Phi_0)$ leads to the condition $B_1 W^2/\Phi_0 \propto W$. By using the software that accompanies figure 15.2, one can see that this is better satisfied as W/L increases.

Also noticeable is that the dependence $\sin(\Delta\varphi_B)$ gets closer and closer to the function $\sin(2\pi y/W)$ as one approaches the limit $W/L \gg 1$. An interesting interpretation follows. Recalling again the 1D limit considered in chapter 3, the linear relation

$$\Delta\varphi_B(y) = \frac{2\pi}{W}y \qquad (15.17)$$

is the expected limiting behaviour. In the current 2D situation, the stray fields cause a distortion in the flux pattern that is increasingly noticeable as the strip becomes narrower and narrower $(W/L \to 0)$. Physically, the screening current pattern within the superconducting islands is increasingly distorted by end effects as W/L decreases. The reader is invited to check this effect with the help of the accompanying scripts.

Figures 15.1 and 15.2 have been generated with the codes

- `figure_15_1.m`
- `Inv_phase_diff.m` (implementation of equations (15.13) and (15.14))
- `Ic_B_strip.m` (implementation of equation (15.16)).

15.2 Resistive transitions in the shunted model

A rather striking feature of Josephson junctions used as circuit elements is the appearance of hysteresis effects in the $V(I)$ characteristics. When the critical current threshold of the junction is overpassed one finds, for instance, that if the transport current is ramped upwards and then downwards one may obtain unequal values of the voltage for a given value of the current. In this section, we present a minimal modelling that will allow one to get familiar with these interesting effects.

To start with, one has to recall that, in order to formulate properly the physics of the problem, the model must include some additional parameters. Up to this point, being focused on the zero voltage properties, only the idealised junction's *critical current I_c* was considered. This quantity represents the capability of superelectrons tunnelling across a certain barrier and characterises the junction. However, a meaningful approximation for the finite voltage states requires further elements. As thoroughly explained by numerous textbooks [1, 2, 3], two main elements have to be considered: (i) the participation of normal electrons (excited tunnelling quasi-particles) and (ii) the capacitive effects related to the accumulation of charge in the opposing electrodes of the junction. Although, more or less sophisticated models are available, many characteristics of the voltage transition may be captured by the simple assumption that is sketched in figure 15.3. To start with, one may assume a shunted model with three constant parameters: I_c, R and C.[7]

15.2.1 The shunted RCSJ model

Let us consider the circuit element shown in figure 15.3 driven by some source connected to its terminals. The full current provided by the source I will be the addition of three contributions:

$$I = \frac{V}{R} + C\frac{dV}{dt} + I_s \tag{15.18}$$

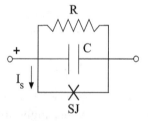

Figure 15.3. Minimal shunted model for the Josephson junction as a circuit element, including resistive and capacitive effects. The 'ideal' non-dissipative Superconducting Junction is represented by its conventional symbol.

---

[7] As a first approximation one typically uses the normal state resistance R_n and estimates C with the plane capacitor expression.

with V being the voltage drop between terminals. The superconducting part obeys the equations (Josephson relations: (3.29) and (3.31))

$$\begin{cases} I_s = I_c \sin(\Delta\varphi) \\ V = \dfrac{\Phi_0}{2\pi}\dfrac{d(\Delta\varphi)}{dt} \end{cases} \qquad (15.19)$$

It is common practice to transform (15.18) into a second order differential equation for $\Delta\varphi$ (replacing V and I_s from (15.19)). Here, we prefer to set the problem in the form of a system of coupled first order equations, that will be eventually solved with the help of MATLAB built-in functions. We proceed as follows. To start with, we re-write the above equations in the dimensionless form

$$\begin{cases} \dfrac{d(\Delta\varphi)}{d\tau} = v \\ \dfrac{dv}{d\tau} = \alpha[\jmath - v - \sin(\Delta\varphi)] \end{cases} \qquad (15.20)$$

where, we have used the definitions: $v \equiv V/I_cR$, $\jmath \equiv I/I_c$, $\tau \equiv t/t_\jmath$, $\alpha \equiv t_\jmath/RC$, with the characteristic time constant for the Josephson junction

$$t_\jmath = \frac{\Phi_0}{2\pi I_c R} \qquad (15.21)$$

Notice that the above system (15.20) allows to solve for the time dependent functions $v(t)$ and $\Delta\varphi(t)$ if one supplies initial conditions $v(t = 0)$ and $\Delta\varphi(t = 0)$. In the forthcoming sections, we describe the application of the related MATLAB scripts, together with the physical discussion on the arising $v(\jmath)$ relations.

15.2.2 $v(\jmath)$ relations: overdamped regime

The numerical solution of equations (15.20) may be achieved for any value of the characteristic parameter α, but it will be instructive to analyse and compare different regimes, as classified through this parameter. To start with, we concentrate on the so-called *overdamped* regime, i.e. for large values of α. Mathematically, the term in brackets must be very small ($[\ldots] \approx 0$). Physically, notice that $\alpha \gg 1$ implies that the RC characteristic time (charging of the capacitor) is very small as compared to the junction's $t_\jmath$. As a consequence, the phase variations will be the dominant factor and dissipation weigh more than capacitance. A closed form solution of the $v(\jmath)$ relation may be obtained with relative ease under such conditions. Thus, by using $[\ldots] \to 0$ one gets

$$\frac{d(\Delta\varphi)}{d\tau} = \jmath - \sin(\Delta\varphi) \qquad (15.22)$$

and may proceed as follows. Consider that a certain level of current $\jmath$ is established by the source. The above equation allows us to obtain $\Delta\varphi(\tau)$ by integration

$$\int d\tau = \int \frac{d(\Delta\varphi)}{J - \sin(\Delta\varphi)}. \tag{15.23}$$

This gives[8]

$$\tau = \frac{-2}{\sqrt{J^2 - 1}} \tan^{-1}\left[\frac{1 - J\tan(\Delta\varphi/2)}{\sqrt{J^2 - 1}}\right] \tag{15.24}$$

which implies

$$\Delta\varphi = 2\tan^{-1}\left[\frac{1}{J} + \sqrt{1 - \frac{1}{J^2}}\,\tan\left(\frac{\tau}{2}\sqrt{J^2 - 1}\right)\right] \tag{15.25}$$

It is apparent that $\Delta\varphi$ will display the periodicity of the tangent function, i.e. it will exhibit the period

$$\tau_{\text{over}} = \frac{2\pi}{\sqrt{J^2 - 1}} \tag{15.26}$$

or, in physical units

$$t_{\text{over}} = \frac{2\pi t_J}{\sqrt{J^2 - 1}} \tag{15.27}$$

Eventually, the stationary $v(J)$ relation is obtained by averaging over a period

$$\langle v \rangle = \frac{1}{\tau}\int_0^\tau v\,d\tau = \frac{1}{\tau}\int_0^{2\pi} d(\Delta\varphi) \tag{15.28}$$

and, straightforwardly

$$\langle v \rangle = \sqrt{J^2 - 1} \tag{15.29}$$

This relation represents the limiting behaviour of a junction for which the capacitive effect is fully negligible, and applies to the range $J \geqslant 1$. Now, the reader may wonder about what occurs if C is small but not that small. Also, in passing one may recall that equation (15.20) admits the solution $\langle v \rangle = 0$ in the region of small applied current $J^2 < 1$ (in fact, having $J = \sin(\Delta\varphi) = \text{constant}$ as expected). In order to provide a means of acquiring intuition and knowledge on all these facts, below we present the numerical solution of equation (15.20). With this, one can explore the behaviour of the junctions for arbitrary values of α and for the full range of values of the current $0 < J < \infty$. As an example of what can be done, figure 15.4 shows the results for the 'overdamped' case $\alpha = 20$. It is apparent that this perfectly follows the analytical approximation for $\alpha \gg 1$, and that no hysteresis is observed. It has been created with the help of the following codes

---

[8] As the reader may straightforwardly check, for example, by using MATHEMATICA. Here the integration constant is chosen to be zero. In fact, the actual value of the constant is not relevant for our eventual purposes, i.e. averaging over a period.

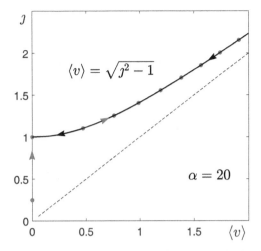

Figure 15.4. $\langle v \rangle (j)$ characteristic for an overdamped junction ($\alpha = 20$) as calculated by numerical solution of equation (15.20). The arrows label the upwards (red symbols)/downwards (black line) solutions along the cycle ($j: 0 \rightarrow 2.5 \rightarrow 0$) that are coincident. In this case, the function $\langle v \rangle = \sqrt{j^2 - 1}$ overlaps accurately on the numerical solution. Also shown is the high current asymptote $\langle v \rangle = j$. 🔗. Source codes available at https://iopscience.iop.org/book/978-0-7503-2711-4.

- VI_JJ_overdamped.m (main)
- diffeq_JJ.m
- two_max.m
- createfigure_overdamped.m

The core of the calculations is the numerical solution of the system (15.20). Such equations, defined by the function diffeq_JJ() are solved with MATLAB built-in functions through the main program. The 'averaged' current voltage characteristic is obtained along the lines described above, i.e. a given value of j_0 is established, then an eventual stationary oscillating regime $v(\tau; j_0)$ occurs, and one integrates in time so as to obtain $\langle v(j_0) \rangle$. In order to ensure correct performance, we proceed by finding two trusted maxima[9] of the oscillating $v(\tau)$ and then integrating in between. The process may be iterated for any desired range of values of j_0. In our case, the (dimensionless) current has been ramped up from 0 to a maximum value 2.5 and then ramped down to 0 again. With the aim of accounting for possible history dependent phenomena the initial values of v and $\Delta \varphi$ at each value of the imposed current density are inherited from the preceding stationary regime. As expected, no hysteresis has been observed.

Considering the complexity of the system under study, it is of interest to analyse in some detail the full response (transient/stationary) of the functions $v(\tau)$ and $\Delta \varphi(\tau)$, successive to the application of some current step. As an example of what can be done, we include figure 15.5 that illustrates the establishment of the stationary regime under different conditions. To the left, we show the time dependence for two

---

[9] One needs to take care because in the transient regime spurious maxima occur.

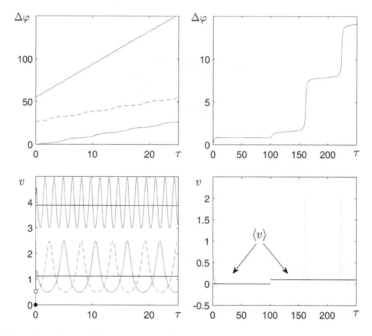

Figure 15.5. Time dependence of the voltage and phase variation across the Josephson junction for the RCSJ model as calculated by numerically solving equation (15.20) for the overdamped case $\alpha = 20$. To the left, we plot the response of the junction to the applied currents $\jmath = 1.5$ (lower lines) and $\jmath = 4$ (upper line). The dashed curves (case $\jmath = 1.5$) correspond to application of the current density starting from a different initial phase difference. To the right, we show the sudden transition from the state given by $\jmath = 0.75$ to the state $\jmath = 1.005$ at $\tau = 100$. Average voltages $\langle v(\jmath_0)\rangle$ are displayed as horizontal lines. ⬀. Source codes available at https://iopscience.iop.org/book/978-0-7503-2711-4.

different values of the applied current ($\jmath = 1.5$, 4) and for different initial conditions $v(\tau = 0)$ and $\Delta\varphi(\tau = 0)$. It is apparent that the stationary regime is established quickly (in units of the characteristic time $t_\jmath$) and that the average voltage $\langle v(\jmath_0)\rangle$ does not visibly depend on the initial conditions. This relates to the observation that the $\langle v\rangle(\jmath)$ curves do not display hysteresis effects at all. To the right of figure 15.5, we plot the transition between an initial non-dissipative state ($\jmath = 0.75$) to the dissipative regime ($\jmath = 1.005$). On the one side, one may clearly identify the feature of superconducting tunnelling ($\Delta\varphi = $ constant, $\langle v\rangle = 0$). On the other side, for $\jmath$ slightly above the critical value ($\jmath \gtrsim 1$), dissipation may be interpreted as follows. $\langle v\rangle$ appears as an average over a series of pulses that occur when $\Delta\varphi$ suffers a 'slippage' of 2π after remaining practically constant. The quantum phase slip by 2π is known as the elementary resistive process in Josephson junctions. Figure 15.5 has been created with the help of the following codes

- `Vt_overdamped.m` (main)
- `diffeq_JJ.m`
- `two_max.m`
- `upper_left.m`, `lower_left.m`, `upper_right.m`, `lower_right.m`

15.2.3 $v(\jmath)$ relations: underdamped regime

In order to highlight the features of underdamped junctions ($\alpha \ll 1$), in this section we analyse the numerical solution of equation (15.20) for the case $\alpha = 0.25$. Figure 15.6 displays the results for $\langle v \rangle$ corresponding to the process

$$\jmath: 0 \to 2.5 \to 0 \tag{15.30}$$

As before, for a given step in the applied current, both v and $\Delta \varphi$ undergo a transient period, and $\langle v \rangle$ has to be carefully defined within the stationary regime. Here, owing to the relevance of capacitive effects, and thus of the RC characteristic time, one may observe new phenomena superimposed on the intrinsic superconducting tunnelling. To start with, transient effects will be noticeable. Related to this, the importance of the initial conditions for the solution of equation (15.20) is manifest, and accompanied by outstanding hysteresis effects. Thus, starting from the state $\jmath = 0$ one finds the expected superconducting behaviour, i.e. $\langle v \rangle = 0$ for $\jmath \leqslant 1$. Then, a sudden jump to the ohmic behaviour $\langle v \rangle = \jmath$ occurs. Outstandingly, when the applied current is reverted back, a noticeable hysteresis displays. In the case under consideration, zero voltage is not reached until $\jmath$ goes down to the characteristic value $\jmath^* \approx 0.6$.

As the reader may check, by changing parameters in the accompanying software, the specific hysteresis 'cycle' followed by $\langle v \rangle (\jmath)$ depends on the exact value of the parameter α. By using a thermodynamic language, the competence between the characteristic time scales $t_\jmath$ and RC determines for which of the possible values $\jmath^*$

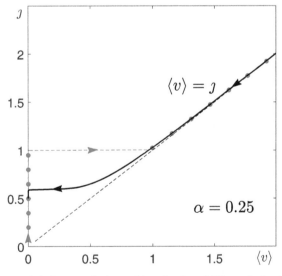

Figure 15.6. $\langle v \rangle (\jmath)$ characteristic for an underdamped junction ($\alpha = 0.25$) as calculated by numerical solution of equation (15.20). The arrows label the upwards (red symbols)/downwards (black line) solutions along the cycle ($\jmath: 0 \to 2.5 \to 0$). The hysteresis effect is apparent. Also shown is the high current asymptote $\langle v \rangle = \jmath$ that is reached for $\jmath \gtrsim 1.5$ in this case. ✑. Source codes available at https://iopscience.iop.org/book/978-0-7503-2711-4.

one reaches the stationary regime $\langle v \rangle = 0$. Clearly, the limiting cases $(\alpha \gg 1)$ and $(\alpha \ll 1)$ correspond respectively to $j^* = 1$ and $j^* = 0$.

Figure 15.6 has been created with the help of the following codes

- `VI_JJ_underdamped.m` (main)
- `diffeq_JJ.m`
- `two_max.m`
- `createfigure_underdamped.m`

Indeed, these are nothing but the customised versions of the corresponding ones for the overdamped case, but are given separately for the users' benefit.

Again, it will be useful to analyse the full behaviour of the functions $v(\tau)$ and $\Delta\varphi(\tau)$, characterising the response to an external current source. Below, we show the results obtained by solving equation (15.20) under the same processes studied for the overdamped case, but now applied to the underdamped condition $\alpha = 0.25$. As said above, the explicit solution of the time regime, reveals the importance of capacitive effects for such kind of junctions.

Following the same flow chart as before, to the left of figure 15.7 we show the time dependence for applied currents $j = 1.5$, 4, and for different initial values $v(\tau = 0)$

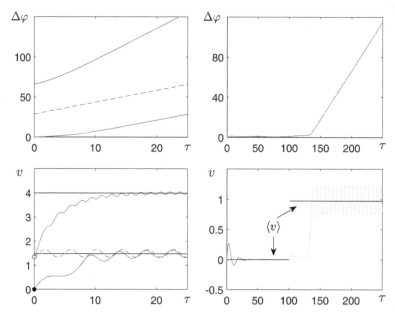

Figure 15.7. Time dependence of the voltage and phase variation across the Josephson junction for the RCSJ model as calculated by numerically solving equation (15.20) for the underdamped case $\alpha = 0.25$. To the left, we plot the response of the junction to the applied currents $j = 1.5$ (lower lines) and $j = 4$ (upper line). The dashed curves (case $j = 1.5$) correspond to application of the current density starting from a different initial phase difference. To the right, we show the sudden transition form the state given by $j = 0.75$ to the state $j = 1.005$ at $\tau = 100$. Average stationary voltages $\langle v(j_0) \rangle$ are displayed as horizontal lines. $\mathcal{O}$. Source codes available at https://iopscience.iop.org/book/978-0-7503-2711-4.

and $\Delta\varphi(\tau = 0)$. In this case, the influence of the time constant RC is apparent. Notice that for the case $\jmath = 1.5$ one has a totally different behaviour of $v(\tau)$ depending on the initial condition. Also manifest is the sharpness of the transition from the superconducting regime ($\langle v \rangle = 0$) to the 'ohmic' behaviour ($\langle v \rangle = \jmath$) when $I = I_c$ (equivalently $\jmath = 1$) is reached from below. This is illustrated in the right-side panels that show the transition from $\jmath = 0.75$ to $\jmath = 1.005$.

Figure 15.7 has been created with the help of the following codes

- Vt_underdamped.m (main)
- diffeq_JJ.m
- two_max.m
- upper_left.m, lower_left.m, upper_right.m, lower_right.m

15.3 Review exercises and challenges

15.3.1 Review exercises

Review exercise R15-1

An alternative way to obtain the stream function (equation (15.13)) is to start with Ginzburg–Landau's equation (equation (15.2)) so as to derive the superconducting phase θ and subsequently obtain **K**. This may be done as follows. Notice that if one uses a gauge selection such that $\nabla \cdot \mathbf{A} = 0$, the stationary regime condition for the current $\nabla \cdot \mathbf{K} = 0$ leads to

$$\frac{\partial^2\theta}{\partial x^2} + \frac{\partial^2\theta}{\partial y^2} = 0, \tag{15.31}$$

i.e. $\theta(x, y)$ satisfies Laplace's equation. Take $\mathbf{A} = -By\hat{\imath}$, and by identifying the components of **K** with the partial derivatives of θ, solve Laplace's equation for θ, and then derive **K**.

Hint: consider the boundary conditions for $\partial_x\theta$ and $\partial_y\theta$ along the perimeter.

Review exercise R15-2

The maximum superconducting current (critical current, I_c) in equation (15.15) may be found by treating $\Delta\varphi_0$ as a variable, and recalling well known trigonometric relations. Show that the following relation holds for average values (along y in this case) of the gauge invariant phase difference

$$\max\langle\sin(\Delta\varphi_0 + \Delta\varphi_B)\rangle = \sqrt{\langle\sin(\Delta\varphi_B)\rangle^2 + \langle\cos(\Delta\varphi_B)\rangle^2} \tag{15.32}$$

In the case under consideration, as related to the symmetry along y axis one has $\langle\sin(\Delta\varphi_B)\rangle = 0$ and, when the maximum is reached, this leads to $\cos(\Delta\varphi_0) = 0$. Then, equation (15.16) follows immediately

$$\max\langle\sin(\Delta\varphi_0 + \Delta\varphi_B)\rangle = |\langle\cos(\Delta\varphi_B)\rangle| \tag{15.33}$$

Recall that the physical counterpart of the above mathematical treatment is as follows. The maximum current flow across the junction is allowed when the wave function within the superconductors adjusts so as to fulfil the condition $\sin(\Delta\varphi_0) = \pm\pi/2$, and takes the value given by (15.16).

Review exercise R15-3

Figure 15.7 illustrates the origin of the sharp transition from the superconducting to the ohmic regime when the applied current density increases from $I = 0.75I_c$ to $I = 1.005I_c$ in an underdamped junction. A sudden jump occurs from the zero voltage regime $v = 0$ to a stationary situation centred at $\langle v \rangle \lesssim 1$. On the other hand, as one may deduce from figure 15.6, when the applied current is reverted, the situation is different, i.e. the $\langle v \rangle$ curve is hysteretic.

Show that, going from the resistive state $I = 1.005I_c$ back to $I = 0.75I_c$ leads to a stationary situation with $\langle v \rangle \neq 0$. Make a plot akin to the right-side panels of figure 15.7 to show it.

Hint: customise the script Vt_underdamped.m. The eventual stationary state is given by $I = 0.75I_c$ and $\langle v \rangle = 0.67$.

15.3.2 Challenges

Exercise C15-1

The presence of Abrikosov[10] vortices in the proximity of Josephson junctions introduces noticeable features in the $I_c(B)$ dependence (figure 15.8). The 'classical' diffraction pattern losses symmetry and can even display minima where one would expect maxima. This is illustrated in the plot. In brief, as the vortex carries a variation of 2π of the superconducting phase, placing it close to the junction allows to control the transport properties under an applied magnetic field (equation (3.29)). As shown by Clem [5], the basic experimental features may be explained by a model that relies on the additive nature of the physical quantities involved. More specifically, for the gauge invariant phase difference one may write

$$\Delta\varphi = \Delta\varphi_B + \Delta\varphi_V \qquad (15.34)$$

with $\Delta\varphi_B$ being the phase variation related to the applied magnetic field (section 15.1) and $\Delta\varphi_V$ the phase variation related to the presence of the vortex.

---

[10] Strictly speaking, this problem refers to a 2D situation, in which case one should more properly speak about 'Pearl' vortices, whose precise mathematical expressions differ. Nevertheless, what is relevant here is to distinguish them (they are equivalent in this respect) from the inherent Josephson vortices in the case of a junction.

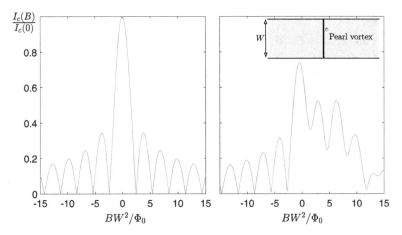

Figure 15.8. Influence of a Pearl vortex close to the Josephson junction between two semi-infinite strips of width W. To the left one has the expected mirror symmetric 'interference pattern' for the plain junction. The right panel shows the deformed critical current pattern in the presence of the vortex. $I_c(0)$ is the critical current in the absence of either applied field or pinned vortices.

In this problem, we propose to investigate the influence of the position of the vortex, its multiplicity, its orientation relative to the actual applied magnetic field ($\pm$), etc by taking advantage of the resources introduced in section 15.1, here applied to the case of a long sample ($L \gg W$). In particular, we suggest to modify the scripts `Ic_B_strip.m` and `Inv_phase_diff_strip.m`. By this means, and by changing the position of the vortex one may produce such plots as the example in figure 15.8, that corresponds to the presence of a vortex at the position $(x_v, y_v) = (0.05W, 0.2W)$. The origin of coordinates was taken at the centre of the junction.

What is the effect of changing the sign of vorticity? What occurs by changing y_v to $-y_v$?

Hint: first, one may show that the invariant phase difference related to the superconducting currents induced by the applied field for a long strip of width W may be obtained from the expression (15.14). Check that one gets

$$\Delta\varphi_B(y) = \Delta\varphi_0 + \frac{16BW^2}{\pi^2\Phi_0} \sum_{n=0}^{\infty} \frac{(-1)^n \sin[(2n+1)\pi y/W]}{(2n+1)^3} \qquad (15.35)$$

On the other hand, as recalled in the work by Clem, the invariant phase difference related to the presence of the vortex $\Delta\varphi_v$ may be obtained by standard conformal mapping techniques because φ_v satisfies Laplace's equation. Taking advantage of this property one obtains the following expression for the phase variation induced across the junction

$$\Delta\varphi_v(y) = -2\tan^{-1}\left[\frac{\sin(\pi y/W) - \cosh(\pi x_v/W)\sin(\pi y_v/W)}{\sinh(\pi x_v/W)\cos(\pi y_v/W)}\right] \qquad (15.36)$$

References

[1] Barone A and Paternò G 1982 *Physics and Applications of the Josephson Effect* (New York: Wiley)

[2] Tinkham M 1996 *Introduction to Superconductivity International Series in Pure and Applied Physics* (New York: McGraw-Hill)

[3] Orlando T P and Delin K A 1991 *Foundations of Applied Superconductivity* (Reading, MA: Addison-Wesley)

[4] Clem J R 2010 Josephson junctions in thin and narrow rectangular superconducting strips *Phys. Rev.* B **81** 144515

[5] Clem J R 2011 Effect of nearby pearl vortices upon the I_c versus b characteristics of planar Josephson junctions in thin and narrow superconducting strips *Phys. Rev.* B **84** 134502

Part IV

Source codes

Appendix A

- **FIGURE 7.3** (DERIVING ACTIVATION ENERGY (γ) VERSUS TEMPERATURE)
```
R_data_inv.m (main)
T_resistivity3.m
inverter.m
createfigure_recover.m
```

- **FIGURE 7.4** (V(I) CURVES FOR DIFFERENT VALUES OF γ)
```
V_I_curves_AH.m (main)
f_AH.m
inv_f_AH.m
int_inv_f.m
interp_int_inv.m
int_combined.m
createfigure_V_I.m
```

- **FIGURE 7.5** (INVERSION OF $R(T)$ DATA TO DERIVE γ)
```
get_data_r.m
R_data_inv_r.m (main)
inverter_r.m
createfigure_recover_r.m
R_YBCO.dat
```

Chapter 8: Flux transport in type-II superconductors

In order to ease the practical application, together with the codes that allow to create data files, the readers are also provided with data structures that allow to straightforwardly create the figures of chapter 8. Some of them are just a single data file (.dat) that should be accessible at the working directory. In other cases, we provide a .zip compressed file. The user must unzip to the working directory and keep the suggested name of the destination folder. This will automatically recreate any internal subdirectory structure. For instance, within the MATLAB command line one should type something like

```
>> unzip('zipped_data_fig_8_10.zip','data_fig_8_10')
```

- **FIGURE 8.3** (TRANSPORT CURRENT IN THE CS FOR SLAB GEOMETRY)
```
slab_cs_qp_transport.m (main)
matrixMXt.m
selfXt.m
mutualXt.m
plot_profile.m
subplot_current.m
subplot_field.m
zipped_data_fig_8_3.zip
```

- **FIGURE 8.5** (FLUX LINES IN THE CS FOR A FINITE CYLINDER)
```
cyl_cs_qp.m (main)
generate_grid.m
matrixM.m
self.m
mutual.m
psi_pot_unif.m
profile_Jcyl_lines.m
matrixM_augmented.m
cylinder7.dat
```

- **FIGURE 8.6** (MAGNETISATION LOOPS IN THE CS FOR A FINITE CYLINDER)
```
cyl_cs_qp.m (main)
generate_grid.m
matrixM.m
self.m
mutual.m
psi_pot_unif.m
magnetization_loops.m
e_cylinder.m
zipped_data_fig_8_6.zip
```

- **FIGURE 8.7** (FLUX LINES IN A SUPERCONDUCTOR/PERMANENT MAGNET ARRANGEMENT)
```
cyl_cs_lev_qp.m (main)
generate_grid.m
matrixM.m
self.m
mutual.m
psi_pot_magnet.m
profile_Jcyl_lines_magnet.m
matrixM_augmented_magnet.m
cylinder68.dat
```

- **FIGURE 8.9** (RELAXATION OF THE CURRENT PENETRATION PROFILE IN A TYPE-II SUPERCONDUCTOR: PIECE-WISE APPROXIMATION)
```
slab_PW_qp_transport.m (main)
matrixMXt.m
selfXt.m
mutualXt.m
relax_PW_profile.m
createfigure_PW.m
zipped_data_fig_8_9.zip
```

- FIGURE **8.10** (RELAXATION OF THE CURRENT PENETRATION PROFILE IN A TYPE-II
  SUPERCONDUCTOR: POWER-LAW MODEL)
  ```
  slab_PL_fmincon_transport.m (main)
  cost.m
  transport.m (only for OCTAVE users)
  matrixMXt.m
  selfXt.m
  mutualXt.m
  relax_PL_profile.m
  createfigure_PL.m
  zipped_data_fig_8_10.zip
  ```

Chapter 9: Shape effects: demagnetizing fields

- FIGURES **9.2**, **9.3**, AND **9.4** (MAGNETIC FIELD LINES AND MAGNETISATION IN
  ELLIPSOIDS AND CYLINDERS: MEISSNER STATE)
  ```
  ellip_meiss.m (main script, ellpsoids)
  cyl_meiss.m (main script cylinders)
  generate_grid_surf_ellip.m
  generate_grid_surf.m
  self.m
  mutual.m
  matrixM.m
  psi_pot_unif.m
  psi_pot_SC.m
  plot_meiss_ellips.m
  plot_meiss_cyl.m
  z_ellipsoid.m
  e_cylinder.m
  Magnetization_field.m
  ```

- FIGURES **9.6**, **9.7** (MAGNETIC FIELD LINES AND MAGNETISATION IN ELLIPSOIDS AND
  CYLINDERS: CRITICAL STATE)
  ```
  ellip_cs_qp.m (main script, ellpsoids)
  cyl_cs_qp.m (main script cylinders)
  generate_grid_ellip_sym_areas.m
  generate_grid.m
  self.m
  mutual.m
  matrixM.m
  psi_pot_unif.m
  psi_pot_ellips_SC_areas.m
  psi_pot_cylinder_SC_areas.m
  plot_cs_ellip.m
  plot_cs_cylinder.m
  ```

```
Magnetization_loop_ellip.m
Magnetization_loop_cylinder.m
```

- FIGURE **9.8** (MAGNETISATION OF ELLIPSOIDS AND CYLINDERS: EXTREME GEOMETRIES)
```
bean_model.m
mikheenko_model.m
figure_9_8.m
```

- FIGURE **9.9** (FIELD PENETRATION IN FLAT SUPERCONDUCTORS)
```
figure_9_9.m
```

Chapter 10: Thin superconductors: the stream function method

The optimization MATLAB codes issued for this chapter have been successfully tested against other minimizers (specifically, the FORTRAN package LANCELOT [1]). Nonlinear optimization scripts in OCTAVE v5.1.0 were tested without success by this author.

- FIGURE **10.2** (STREAM FUNCTION FOR A SQUARE PLATELET)
```
figure_10_2.m
G_example.dat
```

- FIGURES **10.5, 10.6, 10.7,** AND **10.9** (CURRENT DENSITY STREAMLINES AND MAGNETIC FLUX PATTERN FOR ISOTROPIC/ANISOTROPIC PLATELETS)
```
h_perp_plate.m (main script, perpendicular field)
h_incl_plate.m (main script inclined field)
geom_plate.m
self_M.m
mutual_M.m
Q_from_M.m
Kritical_M.m
Kritical_inclined.m
cost.m
figure_10_5.m
magnetization.m
figure_10_6.m
Bz_profile.m
figure_10_7.m
Bz_profile_inclined.m
figure_10_9.m
```

Chapter 11: Magneto-optical imaging of superconductors

- FIGURES **11.2, 11.3, 11.4,** AND **11.5** (CURRENT DENSITY STREAMLINES AND MAGNETIC FIELD PATTERN AROUND SUPERCONDUCTING BEAMS IN THE MEISSNER STATE)

```
get_m.m
inverter_brandt.m
get_Sx_Sz.m
inverter_x.m
inverter_z.m
figure_11_2.m
figure_11_3.m
read_files.m
figure_11_4.m
integrate_field.m
figure_11_5.m
```

- FIGURES **11.6** AND **11.7** (CURRENT DENSITY STREAMLINES AND MAGNETIC FIELD PATTERN AROUND SUPERCONDUCTING BEAMS IN THE CRITICAL STATE)

```
brandt_indenbom_model.m
recover_K.m
```

- FIGURE **11.8** (CURRENT DENSITY STREAMLINES AND MAGNETIC FIELD PATTERN AROUND SUPERCONDUCTING PLATELETS IN THE CRITICAL STATE)

```
MOI_platelet.m
figure_10_7_like.m
```

Chapter 12: Interaction with magnets: force microscopies

- FIGURE **12.2** (POINT DIPOLE ABOVE A SUPERCONDUCTING FILM WITH DIFFERENT VALUES OF λ_L)

```
figure_12_2.m
incident_exact.m
reflected.m
penetrating.m
transmitted.m
m_point.m
```

- FIGURE **12.3** (REPULSION FORCE BETWEEN A POINT DIPOLE AND A SUPERCONDUCTING FILM FOR DIFFERENT VALUES OF λ_L/thickness)

```
force_dipole_mfm.m
```

- FIGURE **12.4** (REPULSION FORCE BETWEEN A CONICAL MFM TIP AND A SUPERCONDUCTING FILM FOR DIFFERENT VALUES OF λ_L/thickness)

```
figure_12_4.m
force_mfm_cone.m
plot_cone.m
```

- FIGURE **12.5** (RECOVERY OF λ_L BY INVERSION OF MFM DATA)
  ```
  force_mfm_recover.m
  ```

- FIGURE **12.6** (RECOVERY OF λ_L BY INVERSION OF MFM DATA: INFLUENCE OF THE SAMPLE THICKNESS)
  ```
  force_mfm_recover_b.m
  ```

Chapter 13: Interaction with magnets: levitation

- FIGURE **13.2** (LEVITATION FORCE: DIPOLE–DIPOLE APPROXIMATION)
  ```
  plot_dipole_dipole.m (main)
  integ_F.m
  integ_K.m
  exact_cyl_g.m
  vec_pot_dip.m
  create_figure_dipole_dipole.m
  ```

- FIGURE **13.3** (LEVITATION FORCE: FINITE MAGNET APPROXIMATION)
  ```
  plot_semiinfinite.m (main)
  semiinfinite_sc.m
  integ_F.m
  integ_K.m
  exact_cyl_g.m
  create_figure_semiinfinite.m
  ```

- FIGURE **13.4** (LEVITATION FORCE: 'UNIFORM' FIELD APPROXIMATION)
  ```
  plot_uniform.m (main)
  uniform_field.m
  integ_F.m
  integ_K.m
  exact_cyl_g.m
  create_figure_uniform.m
  ```

- FIGURE **13.5** (SURFACE CURRENT DISTRIBUTION ON THE SUPERCONDUCTOR: ANALYTICAL VERSUS NUMERICAL)
  ```
  cyl_meiss_lev.m (main)
  generate_grid_surf.m
  matrixM.m
  mutual.m
  self.m
  psi_pot_magnet.m
  integ_K.m
  exact_cyl_g.m
  plot_meiss_num.m
  create_figure_meiss_num.m
  ```

- **FIGURE 13.6** (LEVITATION FORCE IN THE MEISSNER STATE: ANALYTICAL VERSUS NUMERICAL)
  ```
  cyl_meiss_lev2.m (main)
  generate_grid_surf.m
  matrixM.m
  mutual.m
  self.m
  psi_pot_magnet.m
  force_profile_meissner.m
  force_magnet.m
  force_2_loops.m
  integ_F.m
  exact_cyl_g_2.m
  plot_force_meiss_num_disk.m
  create_figure_f_num_disk.m
  ```

- **FIGURES 13.7 AND 13.8** (LATERAL STABILITY: MEISSNER STATE VERSUS CRITICAL STATE)
  ```
  tape_meiss_lev.m (main for 13.7)
  tape_cs_lev.m (main for 13.8)¹
  generate_grid_tape.m
  matrixM_tape.m
  mutual_tape.m
  psi_pot_tape.m
  psi_pot_tape_SC.m
  plot_meiss_tape.m (for 13.7)
  create_figure_meiss_tape.m (for 13.7)
  plot_cs_tape.m (for 13.8)
  create_figure_cs_tape.m (for 13.8)
  ```

- **FIGURES 13.9 AND 13.10** (HYSTERESIS IN THE LEVITATION FORCE)
  ```
  cyl_cs_lev_qp.m (main for 13.9)
  cyl_cs_lev_qp_small.m(main for 13.10)
  generate_grid.m
  matrixM.m
  mutual.m
  self.m
  psi_pot_magnet.m
  psi_pot_SC.m
  force_profile_critical_state.m
  force_profile_critical_state.m_upper.m (for 13.10)
  force_profile_critical_state.m_lower.m (for 13.10)
  ```

---

[1] The quadratic program of MATLAB that is used to solve the CS problem gives a warning, but the soluton was successfully checked against alternative solvers, in particular against OCTAVE.

```
force_magnet.m
force_2_loops.m
build_figure_lev_cycle.m (for 13.9)
build_figure_lev_cycles.m (for 13.10)
subfigure_desc.m (for 13.9)
subfigure_asc.m (for 13.9)
subfigure_asc_low.m (for 13.10)
subfigure_asc_cycle.m (for 13.10)
```

- FIGURE **13.11** (HYSTERESIS IN THE (LATERAL) GUIDANCE FORCE)
  ```
  bar_cs_lev_qp_octave.m
  ```
  (main for users of OCTAVE)
  ```
  bar_cs_lev_fmincon_matlab.m
  ```
  (main for users of MATLAB)
  ```
  generate_grid_bar.m
  matrixM_tape.m
  mutual_tape.m
  psi_pot_tape.m
  psi_pot_tape_SC.m
  cost.m
  ```
  (to be used in the main script of MATLAB)
  ```
  build_figure_guidance.m
  small_cycle.m
  medium_cycle.m
  large_cycle.m
  subfigure_decrease.m
  subfigure_increase.m
  ```

Chapter 14: Magnets and superconductors: cloaking devices

- FIGURES **14.1**, **14.2**, AND **14.3** (MAGNETIC SHIELDING: MAGNETS OR SUPERCONDUCTORS. MAGNETIC CLOAKING: MAGNETS AND SUPERCONDUCTORS)
  ```
  cloak_cyl_meiss.m
  ```
  (main script)
  ```
  define_param_meiss.m
  generate_mesh_meiss.m
  B_xy_het_p.m
  B_xy_sc.m
  B_xy_m.m
  B_app_project.m
  matrixM_II_par.m
  mutual.m
  vector_field_plot_A.m
  A_xy_g_sc.m
  A_xy_g_m.m
  ```

- FIGURES **14.4** AND **14.5** (MAGNETIC CLOAKING: MAGNETIC FIELD LINES AND MAGNETIC MOMENT OF THE HETEROSTRUCTURE)
`cloak_cyl.m` (main script)
`define_param.m`
`generate_mesh.m`
`replicate.m`
`B_xy_het_p.m`
`B_xy_sc.m`
`B_xy_m.m`
`B_app_project.m`
`update_magnet.m`
`matrixM_II_par.m`
`matrixM_IJ_par.m`
`mutual.m`
`magnetic_moment.m`
`vector_field_plot_A.m`
`A_xy_g_sc.m`
`A_xy_g_m.m`
`figure_14_4.m`
`partial_2.mat` (example data)

Chapter 15: Intermediate Josephson junctions

- FIGURES **15.1** AND **15.2** (EDGE-TYPE JOSEPHSON JUNCTIONS, $\Delta\varphi(y)$, $I_c(B)$ DEPENDENCES)
`figure_15_1.m`
`Inv_phase_diff.m`
`Ic_B_strip.m`

- FIGURE **15.4** ($\langle v \rangle (j)$ CHARACTERISTICS FOR AN OVERDAMPED JOSEPHSON JUNCTION)
`VI_JJ_overdamped.m` (main)
`diffeq_JJ.m`
`two_max.m`
`createfigure_overdamped.m`

- FIGURE **15.5** (VOLTAGE AND PHASE VARIATIONS FOR AN OVERDAMPED JOSEPHSON JUNCTION)
`Vt_overdamped.m` (main)
`diffeq_JJ.m`
`two_max.m`
`upper_left.m`
`lower_left.m`
`upper_right.m`
`lower_right.m`

- FIGURE **15.6** ($\langle v \rangle (J)$ CHARACTERISTICS FOR AN UNDERDAMPED JOSEPHSON JUNCTION)

```
VI_JJ_underdamped.m (main)
diffeq_JJ.m
two_max.m
createfigure_underdamped.m
```

- FIGURE **15.7** (VOLTAGE AND PHASE VARIATIONS FOR AN UNDERDAMPED JOSEPHSON JUNCTION)

```
Vt_underdamped.m (main)
diffeq_JJ.m
two_max.m
upper_left.m
lower_left.m
upper_right.m
lower_right.m
```

Reference

[1] Conn A R, Gould N I M and Toint P L 1992 *LANCELOT: A Fortran Package for Large-scale Nonlinear Optimization (Release A)* (Heidelberg: Springer)

Part V

Notation

IOP Publishing

Macroscopic Superconducting Phenomena
An interactive guide
Antonio Badía-Majós

Appendix B

Notation

Many equations in this book are expressed in dimensionless units defined by the actual problem, i.e. in terms of characteristic lengths and physical parameters such as critical fields, current densities, etc. In all cases, connection to physical units is done by means of the International System (S. I.). Below, we give a list of symbols used for the physical quantities and characteristic parameters.

It is worth mentioning that we have used upper-case for the electromagnetic fields $\mathbf{B(r)}$, $\mathbf{J(r)}$, $\mathbf{E(r)}$, that are coarse-grained averages at the macroscopic level. Although changing within the superconductors, they are defined with a resolution that smoothes out fluctuations caused by individual entities (i.e. vortices of flux quanta) with field being understood as $\mathbf{B(r)} \equiv \langle \mathbf{b(r}_v) \rangle$[1]. Eventually, our 'macroscopic fields' are connected to the 'observables' through new averages, now over the sample's volume:

$$\mathbf{M}_V \equiv \frac{\langle \mathbf{B(r)} \rangle}{\mu_0} - \mathbf{H}_a$$

Although not strictly necessary in the formulation of the selected problems, we include the magnetic field strength $\mathbf{H}$ as related to the controllable experimental sources, that produce a magnetic field of value $\mu_0 \mathbf{H}_a$.

As many of the problems are solved by numerical methods that are stated in terms of finite element expressions in some grid of points, we define the corresponding collections of values and represent them in the form of vectors, that are written in Dirac's notation so as to distinguish from real space vectors. Thus, we use $\mathbf{B(r)}$ for the three-dimensional magnetic field, and $|\mathbf{B}_x\rangle$, $|\mathbf{B}_y\rangle$, $|\mathbf{B}_y\rangle$ for the collection of values of the respective components at the points of the grid[2].

---

[1] See section 2.6.
[2] See section 5.2.2.

doi:10.1088/978-0-7503-2711-4ch17

Universal constants

Symbol	Quantity
e	Electron's charge
μ_0	Vacuum's magnetic permeability
Φ_0	Flux quantum

Physical quantities and parameters

Symbol	Quantity
$\mathbf{A}$	Magnetic vector potential
$\mathbf{b}$	Magnetic field (mesoscopic)
$\mathbf{B}$	Magnetic field (macroscopic or single length-scale)
$\mathbf{E}$	Electric field
$\mathbf{f}$	Mechanical force (mesoscopic)
$\mathbf{F}$	Mechanical force (macroscopic)
$\mathcal{F}$	Free energy density
$\mathbf{H}_a$	Applied magnetic field strength
H_p	Full penetration field
H_m	Maximum applied field strength
$\mathbf{j}$	Current density (mesoscopic)
$\mathbf{J}$	Current density (macroscopic)
$\mathbf{K}$	Surface current density/sheet current
$\mathbf{K}_{\mathrm{M}}$	Magnetic surface current density
M_{ij}	Mutual inductance matrix elements
$\mathbf{M}_{\mathrm{EQ}}$	Equilibrium magnetisation
M_p	Saturation magnetisation
$\mathbf{M}_{\mathrm{V}}$	Volume magnetisation
$\mathcal{P}$	Power per unit volume
U_{M}	Magnetic energy density
V	Voltage
$\langle v \rangle$	(time) Averaged voltage
$\mathcal{W}$	Dissipated energy
θ	Wave function's phase
σ	Stream function
ϕ	Electrostatic scalar potential
$\boldsymbol{\mu}(m_x, m_y, m_z)$	Magnetic dipole moment (its components)
$\boldsymbol{\psi}$	Reshaped magnetic vector potential
Ψ	Superconducting order parameter
$\boldsymbol{\omega}$	Superconducting current density (zero field)

As customary, bold font is used for vector quantities in three dimensional space and plain style for scalars.

Material parameters

Symbol	Quantity
H_c	Thermodynamic critical field
H_{c1}	Lower critical field
H_{c2}	Upper critical field
R_n	Normal state resistance
γ	Activation energy
κ	Ginzburg–Landau's parameter
λ_L	London's penetration depth
Λ_L	$\mu_0 \lambda_\mathrm{L}^2$
Λ	In thin films λ_L^2/d
ρ	Resistivity (general)
ρ_f	Flux flow resistivity
ρ_n	Normal state resistivity
σ	Conductivity
μ_r	Relative magnetic permeability
ν	Exponent in power law
ξ	Coherence length

Matrix representations

Symbol	Quantity
$|\mathbf{B}\rangle$	Magnetic field
$\mathbf{D}_x, \mathbf{D}_x$	Finite difference operators
$|\mathbf{H}\rangle$	Magnetic field strength
$|\mathbf{J}\rangle\,(J_i)$	Current density (its elements)
$|\mathbf{K}\rangle$	Surface current density/sheet current
$|\mathbf{K}_\mathrm{M}\rangle$	Magnetic surface current density
$\mathbf{m}\,(m_{ij})$	Mutual inductance matrix (its elements)
$|\mathbf{x}\rangle$	x-coordinates array
$\mathbf{X},\mathbf{Y},\mathbf{Z},$	Biot-Savart operators (Chapter 10)
$|\sigma\rangle$	Stream function
$|\psi\rangle$	Magnetic vector potential

CPSIA information can be obtained
at www.ICGtesting.com
Printed in the USA
BVHW010530060122
625571BV00003B/24